An International Nutrition Index

Concept and Analyses of Food Insecurity and Undernutrition at Country Levels

Doris Wiesmann

PETER LANG
Europäischer Verlag der Wissenschaften

Bibliographic Information published by Die Deutsche Bibliothek
Die Deutsche Bibliothek lists this publication in the Deutsche Nationalbibliografie; detailed bibliographic data is available in the internet at <http://dnb.ddb.de>.

Zugl.: Bonn, Univ., Diss., 2002

D 98
ISSN 0948-1338
ISBN 3-631-50475-6
US-ISBN 0-8204-6084-2

www.peterlang.de

For my parents

An International Nutrition Index

DEVELOPMENT ECONOMICS
AND POLICY
Series edited by Franz Heidhues and Joachim von Braun

Vol. 39

PETER LANG
Frankfurt am Main · Berlin · Bern · Bruxelles · New York · Oxford · Wien

Foreword

Food security and nutrition have been long standing issues in the development debate. Hunger and undernutrition seriously affect the health and working capacity of present and future generations. They have huge economic costs, especially if considered in a long-term perspective. Thus, strategies to improve food security and nutrition are not only worth pursuing for their own sake - they are also key elements of poverty reduction. In spite of these well-known linkages, progress in combating hunger and malnutrition has been lagging behind the targets aspired to for decades.

In this study by Doris Wiesmann, the lack of a commonly accepted, comprehensive measure for food and nutrition security is identified as one of the roadblocks on the way to the eradication of hunger and malnutrition. In an attempt to fill this gap, key principles for the construction of international indices are derived from the debate on the Human Development Index (HDI). An international Nutrition Index (NI) is designed in accordance with these guidelines, comprising the percentage of undernourished (estimated by FAO), the prevalence of underweight in children and the under-five mortality rate.

Possible applications of such an index are exemplified by an international ranking of 97 developing countries and nine countries in transition, and by the consideration of trends from 1981 to 1997. The calculation of disaggregated values for China's provinces and India's states demonstrates that the NI can be used at the subnational level as well. Compared to its components, the NI shows quite close correlations with indicators of micronutrient deficiencies (which could not be included in the index due to scanty data availability). A notable characteristic of the NI is its sensitivity to inequality and poverty at the international level. Taking four measures of income poverty as a yardstick, the NI outperforms GNP per capita, the HDI and even its three components in this respect.

The quantitative analysis of cross section data for 94 countries provides insights into the driving forces behind the international variation of the NI. Whereas GNP per capita is by far the most important determinant, political and social indicators also have a significant impact. The NI tends to rise with increases in urbanisation, immunisation rates of children, and utilisation of adequate obstetric care. In poor countries, agriculture can play a key role in improving food supply. High levels of military expenditure and engagement of countries in wars are shown to have a detrimental effect on food security and nutrition, even apart from their potentially negative consequences for macro-economic performance.

The study can serve as a starting point for the creation of an internationally accepted measure for monitoring food security and nutrition. At the same time, a political debate is needed for setting priorities towards more efficient and effective action against hunger and malnutrition in the future. 'Best practices' have been available for a long time, but lack of political will on the part of leaders and lack of political power amongst the poor has hampered their implementation.

Prof. Dr. Franz Heidhues
University of Hohenheim
Stuttgart

Prof. Dr. Joachim von Braun
International Food Policy
Research Institute (IFPRI)
Washington D.C.

Acknowledgements

The present study is the outcome of a research project on international food security and nutrition monitoring at the Center for Development Research. This work has benefited greatly from the support of my colleagues, my friends and my family.

First of all, I would like to thank Prof. Dr. Joachim von Braun for supervising this study, for his patience and his confidence in me. His idea to develop an international Nutrition Index provided the igniting spark for the present dissertation. His intellectual guidance, encouragement and support throughout the study are greatly appreciated. I am also very grateful to Prof. Dr. Dr. Walter Schug for his interest in my work.

Second, I am grateful also to my colleagues at ZEF for their support and the warm and inspiring atmosphere they created. I am particularly indebted to Dr. Torsten Feldbrügge for stimulating discussions and valuable comments throughout the design stage of the Nutrition Index. I thank Dr. Mingzhi Sheng, who was an indispensable aide for the part on China by sharing her rich knowledge of Chinese statistics, especially on food consumption data. Valuable comments on the study by Dr. Matin Qaim, Dr. Assefa Admassie, Dr. Johannes Jütting, Dr. Abay Asfaw, Dr. Daniela Lohlein, and several participants at the doctoral seminar at ZEF are gratefully acknowledged.

Third, I very much appreciate the comments and assistance of researchers from other institutions and universities. Most of all I would like to thank Dr. Susanne Wildner for diligent reading of large parts of the study and for her helpful advice on methodological matters. I am also very grateful to Dr. Guido Lüchters for his comments on this study from a statistical point of view. I am greatly indebted to Monika Blössner from WHO for answering my numerous questions about the Global Database on Child Growth and Malnutrition patiently and proficiently, and to Mark Hereward from UNICEF and Prof. Dr. Mahendra Dev from the Indira Gandhi Institute of Development Research for providing subnational data for China and India. Dr. Rasmus Heltberg from the University of Copenhagen, Prof. Dr. Krawinkel and Dr. Ulrike Rötten from the University of Giessen are thanked for inspiring suggestions. I am grateful to Mrs Maggie Sanderson for excellent English proof-reading of the whole dissertation, and for thought-provoking comments.

Last but not least, I would like to thank my family - first of all my mother Christa Wiesmann, who never hesitated to take over the housework during the times of my absence. I am grateful to my husband Theo Specht for his support and encouragement, and for helpful formatting and printing assistance. I am very thankful for the patience my son Uwe has shown with his ever busy mother.

Table of Contents

List of Tables

Tables in main text

Tables in Boxes

Tables in Appendix

List of Figures

Figures in main text

Figures in Boxes

Figures in Appendix

List of Boxes

Boxes in main text

Boxes in Appendix

List of Acronyms and Abbreviations

ADEC(X)	Average Dietary Energy Content of Category X (in kcal/kg)
AEA	American Economic Association
CM	Child Mortality (expressed by the under-five mortality rate)
CRED	Center for Research on the Epidemiology of Disasters
CUW	Children Underweight
DES	Dietary Energy Supply per capita per day
DES DIFF	Ranking difference: GNP per capita rank minus DES rank
DESC(X)	Dietary Energy Supply from Category X per capita per day (in kcal)
DHS	Demographic and Health Surveys
EDI	Economic Development Institute (of the World Bank)
EIU	Economist Intelligence Unit
FAO	Food and Agriculture Organisation of the United Nations
FAOSTAT	FAO Statistical Databases
FCND	Food Security and Nutrition Division (of IFPRI)
FIVIMS	Food Insecurity and Vulnerability Information and Mapping Systems
FS(X)	Food Supply of Category X per capita per year (in kg)
GDP	Gross Domestic Product
GIEWS	Global Information and Early Warning System on Food and Agriculture
GINI	Gini Coefficient
GNP	Gross National Product
HDI	Human Development Index
HDI DIFF	Ranking difference: GNP per capita rank minus HDI rank
HDI*	Human Development Index calculated without the GDP component
HDI* DIFF	Ranking difference: GNP per capita rank minus HDI* rank
HI	Hunger Index
HIIK	Heidelberger Institut für internationale Konfliktforschung - Heidelberg Institute for International Conflict Research
HIV/AIDS	Human Immunodeficiency Virus/Acquired Immune Deficiency Syndrome
HPI	Human Poverty Index
IAOS	International Association for Official Statistics

ICCIDD	International Council for the Control of Iodine Deficiency Disorders
IDD	Iodine Deficiency Disorders
ICN	International Conference on Nutrition
ICRG	International Country Risk Guide
IFPRI	International Food Policy Research Institute
IMF	International Monetary Fund
IRIS	Center for Institutional Reforms and the Informal Sector
KOSIMO	Konflikt-Simulations-Modell – Conflict Simulation Model
MI	Micronutrient Initiative
NGO	Non-Governmental Organisation
NI	Nutrition Index
NI DIFF	Ranking difference: GNP per capita rank minus NI rank
$NI_{1,2}$	Nutrition Index calculated without the child mortality component
$NI_{1,3}$	Nutrition Index calculated without the child malnutrition component
$NI_{2,3}$	Nutrition Index calculated without the undernourishment component
$NI_{2,3}$ DIFF	Ranking difference: GNP per capita rank minus $NI_{2,3}$ rank
NI-HDI DIFF	Ranking difference: HDI rank minus NI rank
NI-HDI* DIFF	Ranking difference: HDI* rank minus NI rank
OFDA	Office of United States Foreign Disaster Assistance
OLS	Ordinary Least Squares
PHR	Partnerships for Health Reform
PPP$	Purchasing Power Parity Dollars
PQLI	Physical Quality of Life Index
PRS	Political Risk Services
PUN	Percentage of Undernourished
PUN DIFF	Ranking difference: GNP per capita rank minus PUN rank
SADC	Southern African Development Community
SDP	State Domestic Product
SOWC	State of the World's Children
UN	United Nations
UN ACC/SCN	United Nations Advisory Committee on Coordination/Subcommittee on Nutrition
UNAIDS	Joint United Nations Programme on HIV/AIDS

UNDP	United Nations Development Programme
UNICEF	United Nations Children's Fund
UNU	United Nations University
WHO	World Health Organisation
ZEF	Zentrum für Entwicklungsforschung – Center for Development Research

1 Introduction

Access to adequate food and a decent nutritional status are central to human well-being. Hunger and undernutrition are obvious expressions of basic human rights violations. They have many faces, ranging from loss of energy and apathy to increased susceptibility to disease, anthropometric shortfalls, disability and premature death. Hunger and undernutrition can cripple children's physical and intellectual capabilities for a lifetime and affect the health of future generations. Lack of food and malnutrition are both a consequence and a cause of poverty. Therefore, addressing poverty effectively requires improvements in food security and nutrition. The unholy alliance of desperate poverty and food insecurity is believed to breed violent conflict, terrorism, and fanaticism. Furthermore, in the long run, the economic costs of malnutrition are huge: recent estimates suggest a loss to GDP of 3.5% annually (Horton 1999). This implies that after twenty years, GDP would be double in the absence of malnutrition (Haddad 2001). Despite these strong arguments for combating hunger and undernutrition, these conditions are still the plight of millions of people in our times, especially in developing countries.

1.1 Problem Setting

Notable progress has been made in enhancing food security and preventing malnutrition since 1948, when the right to a standard of living adequate for health and well-being - including food and other basic necessities - was stated in the Universal Declaration of Human Rights. However, considering the goal of the United Nations' World Food Conference in 1974 - that no child should go to bed hungry within a decade - the gap between vision and reality is striking: at the end of the 1990s, about 815 million people worldwide were suffering from hunger to the extent that they could not meet their minimum dietary energy requirements (FAO 2001a). About 150 million children under five were malnourished as measured according to their weight for age, and even more – 182 million – showed moderate or severe stunting growth while 51 million were wasted at the turn of the millennium (UN ACC/SCN 2000). About half of the 11.6 million deaths among children under five in 1995 in developing countries are estimated to be associated with malnutrition (Bailey et al. 1998). Around 3 million preschool children in the developing world suffer from eye lesions caused by lack of dietary Vitamin A, and up to 140 million are estimated to have milder subclinical forms of Vitamin A deficiency, which lead to increased morbidity and higher mortality (MI/UNICEF/Tulane University 1998). The dietary iodine supply of roughly one third of the world population is at risk (WHO/UNICEF/ICCIDD 1999). More than 3.5 billion people in developing countries are affected by iron deficiency and its associated anaemia (UNICEF/UNU/WHO/MI 1999). As these figures clearly confirm, hunger and malnutrition are neither minor nor negligible problems in our modern world, given the extent of suffering they entail for the people concerned, and considering the proportion of the world population affected. This raises the question of what has really been done to overcome these fundamental grievances.

During the1990s, the international community repeatedly addressed the largely unsatisfactory global nutrition situation: examples are the World Summit for Children in 1990, the International Conference on Nutrition in 1992, and the World Food Summit in 1996 in Rome. In 1990, the participants at the World Summit for Children organised by UNICEF agreed to reduce child mortality by one third, cut back moderate to severe malnutrition in children by half, the proportion of babies with low birthweight to less than 10% and iron deficiency anaemia in women by one third, and to virtually eradicate iodine and Vitamin A deficiency by the year 2000 (UNICEF 1990a). The Summit thus focussed on nutritional and health outcomes of women and children, but did not set explicit targets for reducing chronic dietary energy deficiencies in the total population or achieving food security for all. Progress has indeed been made with respect to the above-stated goals, but the actual achievements have fallen short of the expectations. Under-five mortality declined by only 14% during the decade after 1990, which is less than half of the 33% aspired to. The prevalence of underweight in children was only cut back by 17% instead of the envisaged reduction by 50%, whereas the elimination of Vitamin A and severe iodine deficiency showed a better track record, though without being fully achieved (UNICEF 2001). The International Nutrition Conference (ICN) convened in 1992 by FAO and WHO stressed determination to end hunger and reduce all kinds of malnutrition. The main goals were the elimination of famines and associated deaths, as well as the eradication of diseases caused by nutritional deficiencies in areas struck by natural or man-made disasters, by the year 2000 (FAO/WHO 1992). As the meeting was a conference and not a summit of heads of states, no binding agreements on quantitative targets were entered into. In retrospect, the goals proved too ambitious to be reached within the short timeframe (von Braun et al. 1998).

The World Food Summit 1996 in Rome was a gathering of national leaders organised by FAO. The heads of government and ministers approved an extensive Plan of Action on the reduction of hunger and malnutrition and the achievement of food security for all, and made a final declaration. However, whereas the World Summit for Children had put strong emphasis on improving health environments, combating infectious diseases and providing special assistance to mothers and children, health and care aspects are largely missing in the Plan of Action and Rome Declaration on World Food Security (FAO 1996a). This can be seen as a step backwards from the insights and suggestions presented at previous international conferences such as the ICN 1992 (Gabbert 2000). The sole quantitative goal set at the World Food Summit in Rome was the reduction by half of the number of undernourished people in developing countries by 2015 (FAO 1996a). According to FAO's latest assessment of the State of World Food Insecurity, the world is not on track to achieve this aim: since the reference period in 1990-92, the annual average decrease in the number of the undernourished has stood at only 6 million instead of the 20 million projected as necessary to reach the target. There has been a slowdown in the decline of undernourishment, and the gap between reductions realised and reductions needed is widening. The decrease in the number of undernourished people would now have to be expanded to 22 million per year to achieve the goal of the Summit, and this is well above the current level of performance (FAO 2001a). The 2020 Vision Initiative of the International Food Policy Research Institute (IFPRI) has proposed a more ambitious target than the World Food Summit, namely food security for all by the year 2020. However, this goal will not be achieved with 'business as usual', as has recently been stated on the 2020 Vision Conference in Bonn (IFPRI 2001).

Why is there no better track record with reference to the aims enshrined in the declarations passed at international summits and conferences? Lack of political will is the most frequently cited reason for failures to achieve the above-stated targets. Contrary to the repeated pledges made at international meetings, governments of developed and developing nations have not fulfilled their commitments. Yet, whilst acknowledging their notable contributions, it must be said that the international food security and nutrition community – the main UN agencies and researchers active in the field – could also have done better in championing the cause of the hungry and malnourished. As the review of international conferences and summits and the variety of their targets illustrate, there has been a lack of coordination and cooperation among relevant agencies. The agendas of the international meetings, e.g., the World Summit for Children contrasted with the World Food Summit, have been strongly influenced by a deeply entrenched institutional as well as disciplinary divide between agricultural experts concerned with food production on the one hand and nutrition and health professionals on the other[1]. This divide is reflected at top levels in the organisations' mandates, in different lead indicators preferred for monitoring and setting quantitative targets, and in development project planning on the ground. It also cuts through the research community and efforts in data collection[2]. Among researchers, there has been a long-lasting dispute whether national indicators of food availability or anthropometric indicators are better suited to monitor food security and target resources to the neediest. Implications for action differ quite a lot depending on the indicator used: if we look at the FAO estimates of undernourishment based on national food availability, things look worst in Sub-Saharan Africa. Taking child malnutrition as a yardstick, South Asia turns out to be the region that deserves utmost attention and assistance. The reasons for this divergence have not yet been fully clarified. Both types of indicators - estimates based on national food supply and anthropometric indicators - have been criticised for proneness to measurement error, limited reliability and lack of comprehensiveness (see Svedberg 1998 and Klasen 1999[3]). The controversial debate about the 'best measure' vividly illustrates that all existing indicators have their virtues and their shortcomings, and that the complex problems of hunger and malnutrition can hardly be captured by a single indicator. The need for a comprehensive measure, however, has been stressed during the last sessions of the IFPRI Vision 2020 Conference in Bonn. An index might be best suited for quantifying and continuously monitoring changes in a country's or region's food security and nutrition situation.

---

[1] This divide is reflected in the long-term discussions on priority setting in the United Nations Advisory Committee on Coordination/Subcommittee on Nutrition (UN ACC/SCN). Regarding the coordination of UN agencies in the field of food security and nutrition, change is on the way with the Food Insecurity and Vulnerability Information Mapping System (FIVIMS) initiative, which could in the long run also contribute to better interdisciplinary cooperation. FIVIMS are networks of systems that assemble, analyse and disseminate information about the problems of food insecurity and vulnerability. FIVIMS aims to raise awareness about food security issues, improve the quality of food security related data and analysis, promote donor collaboration on food security information systems at country level, encourage better action programmes on poverty and hunger, and to improve access to information through networking and sharing (FAO 1999a).

[2] An example is the data collection efforts of the Demographic and Health Surveys (DHS). Whilst providing a wealth of data on demographic characteristics, maternal and child health and nutrition, the DHS include no explicit information on food consumption or food expenditure. Yet from the viewpoint of food security and nutrition, such complementary information would be most useful for a joint data analysis. Correspondingly, income and (food) expenditure surveys or food consumption surveys mostly fail to cover basic anthropometric indicators.

[3] Klasen has demonstrated how small inadequacies in international reference standards for child growth (possibly 1-3 cm) could significantly distort measured chronic malnutrition prevalence if the maximum of the normal distribution of children's anthropometric records is close to the cut-off point.

Beginning with the release of the Human Development Index (HDI) by UNDP in 1990, indices became increasingly popular in the development discussion throughout the 1990s. In their property as single numbers, they are more 'eye-catching' in the policy debate than a collection of various indicators. Publication in international league tables and comparisons with countries' economic development have proved suitable for fostering a sense of competition among nations and for mobilising political will. Despite the technically weak format in which the HDI was initially introduced, it has greatly stimulated development thinking and policy-making. Unfortunately, the HDI fails to represent food security and nutrition adequately, which is exemplified by Reichel (1991): according to the Human Development Report 1990, Ethiopia's HDI is more than double that of Niger indicating that, in the language of UNDP, human development is twice as high in Ethiopia compared with Niger (UNDP 1990). This assessment is in sharp contradiction of the fact that Ethiopia has been plagued by repeated famines from the beginning of the 1970s, with a dietary energy supply per capita per day of only 1749 kcal in 1986, whereas Niger has achieved an average supply of 2492 kcal in 1986 – information which is not considered in the HDI. The GDP per capita in PPP$ as one HDI component being almost identical in both countries, the better position of Ethiopia in the HDI 1990 ranking is solely attributable to its much higher literacy rate. Reichel concludes, not without polemics, that 'you need not be a proponent of Maslow's hierarchy of human needs to imagine that an Ethiopian just dying from hunger will hardly be able to enjoy his or her ability to read and write'[4] (Reichel 1991, p.60). Similar considerations apply to one of the HDI's predecessors, the Physical Quality of Life Index (PQLI), that is composed of the literacy rate, life expectancy at age one and infant mortality (Morris 1979). The PQLI has no components indicating food security and nutrition more directly than the two health indicators do. The Human Poverty Index (HPI), however, includes a child malnutrition indicator besides four other components (UNDP 1997). But the HPI is meant to reflect a broad set of deprivations of basic necessities. The HPI is not specifically tailored to as-

---

[4] A further example which shows that the HDI may be crudely misleading with respect to food security and nutrition is the comparison of Mauritania and Mozambique. Although dietary energy supply in Mozambique is found to be only 1595 kcal per capita per day in 1990, which is reported to be a 'world-wide record of catastrophic undernourishment' (Reichel 1991, p. 60), the HDI value suggests that Mozambique is, due to its higher literacy rate, superior to Mauritania in terms of human development, ignoring that calorie supply is at a much more favourable level in Mauritania at that time (2322 kcal per capita per day, which would be roughly sufficient on the average). Neither the detrimental impact of the civil wars devastating Ethiopia and Mozambique at the end of the 1990s, nor the effects of failed agricultural policies on food supplies are sufficiently reflected by the HDI (Reichel 1991).

sess hunger and undernutrition, thus containing no direct indicator of food availability. Actually, food security and nutrition have not figured very prominently on the development agenda, as is demonstrated by the relative neglect of respective indicators in international indices meant to capture the essence of the 'quality of life' or 'human development'.

It is thus a straightforward exercise to develop an international Nutrition Index able to fulfil the need for a comprehensive monitoring tool and to raise awareness of food security and nutrition. After, or better, in parallel to the designing of the Nutrition Index presented here, a Hunger Index (HI) has been constructed by Douglas Hicks (Bread for the World Institute 2001). The HI originates from the same basic idea as the NI: creating an international index that is able to reflect different aspects of food security and nutrition. However, the HI is subject to serious constraints in applicability. First, the HI relies, among other indicators, on the Gini-Index of consumption expenditure (excluding the Gini of income), which results in very limited data availability: the HI could only be calculated for a small set of 49 developing countries. Second, the use of stunting in children as a partial indicator prevents the identification of long-term trends for countries, because data on this anthropometric indicator have only become broadly available in the 1990s (UN ACC/SCN 1997). Third, and this is more a conceptual matter, premature mortality as the most tragic consequence of hunger and undernutrition has been neglected[5].

1.2 Objectives and Research Questions

Considering the aforementioned lack of a broadly available comprehensive indicator of food security and nutrition, the main objective of the study is to design an international Nutrition Index (NI) that could fill this gap. The combination of several indicators in one index has major advantages. First, an index composed of different, but related indicators is able to integrate several aspects of a multifaceted phenomenon like food security and nutrition, which cannot be adequately captured by a single indicator. Second, the combination of indicators measured independently from each other may reduce the impact of measurement error on the resulting index. Third, in contrast to the multitude of single indicators that are currently used, assembling various complementary indicators in one index facilitates a quick international and intertemporal over-

[5] Though the NI has been developed without prior knowledge of the HI (which was in fact released at a later point in time than the first publication of the NI), more attention has been devoted to these issues during its design, especially regarding data availability.

view[6]. Last but not least, the introduction of an index of food security and nutrition could end the unfruitful dispute over the 'best measure' and might help to bridge the divide among relevant disciplines and institutions in the field. Further objectives of the study are the application of the NI as a monitoring tool to a broad set of developing countries, and its use in comparison with other international indicators. In order to arrive at policy implications, insights into the driving forces for food security and nutrition as measured by the NI are sought.

The main research questions derived from these objectives refer to a creative, a descriptive and an analytical part of the study. The first question addresses the creation of an international Nutrition Index. The second research question aims at describing the empirical findings across countries and over time, as well as within selected large countries. The last two questions and their subquestions are of an analytical nature and deal with the relation of the NI to other international indicators and the main determinants of its variation.

1. How can an international Nutrition Index be constructed in a conceptually and methodologically sound way, taking into account limits of data availability?

2. What are the numerical results of NI calculation, and which disparities and trends do they show for countries and regions (also, if possible, on a subnational level for large countries)?

3. How does the NI relate to other international indicators – is there an added value of calculating it, and, if yes, what new insights does the NI offer? The following subquestions are addressed:
    - How does the NI relate to other indicators of food security and nutrition, is it possibly redundant vis-à-vis its components, and is it able to reflect nutritional information that has not been included, such as indicators of micronutrient deficiencies?
    - Does the NI display sensitivity to poverty and inequality, and how does it perform in this respect compared to its components, to GNP per capita and the Human Development Index?

---

[6] However, an index cannot substitute for more disaggregated assessments of the various specific nutrition problems, such as is carried out by FIVIMS (Food Insecurity and Vulnerability Information Mapping System) on a subnational level.

- Is the NI redundant vis-à-vis the most common development indicators, how can its functional relationship to GNP per capita and the HDI be described, and which countries can be classified as 'outliers', and why?

4. What are the main determinants of NI variation, and how can their significance in the aggregate NI model be traced back to their impact on its components?

The study is structured according to the sequence of these research questions; its organisation is outlined next.

1.3 Organisation of the Study

The six chapters following the introduction are dedicated to answering the above research questions and are completed by the conclusions and the summary.

In Chapter 2, the theory of international comparisons is outlined with reference to the desirable properties and technical requirements of international indicators, and the most frequently cited pros and cons of international indices are highlighted. Special attention is devoted to a measurement concept of international indices, including criteria for the selection of partial indicators, operations carried out with them, and methods of aggregation. This chapter is mainly based on insights from the theoretical and conceptual discussion that has been incited by the release of the Human Development Index in 1990.

Chapter 3 sets out definitions of food security and nutrition as well as the conceptual framework on which the international Nutrition Index is based. The selection of partial indicators is performed in accordance with the requirements pointed out in Chapter 2. Limitations due to insufficient data availability are considered in the choice of indicators.

Chapter 4 deals with the underlying data for NI calculation and the method of aggregation. An estimation model is developed and applied to supplement the data set on undernourishment. Data on child malnutrition are refined for better international comparability, and the under-five mortality data to be used are described. In the absence of yearly information for each of the indicators, reference periods for NI calculation are defined. Principal components analysis is used to derive the weights for aggregation. For different ways of standardisation of the partial indicators, the resulting indices are checked for redundancy vis-à-vis an index version based on non-standardised components, and the sensitivity of weights is tested.

Chapter 5 is devoted to the presentation of an international ranking according to NI values and the illustration of the world nutrition situation in a map. Patterns and recent trends in food security and nutrition are identified and discussed for continents, regions and countries. Subnationally disaggregated NI values are presented for India's states and China's provinces.

Chapter 6 addresses the relation of the NI to other international indicators, starting with its components and other international food security and nutrition indicators. Beside the issue of redundancy, the ability of the NI to reflect the information content of non-included nutritional indicators is examined. The sensitivity of the NI to poverty and inequality is tested by means of quantitative comparison with various distributional and poverty measures. Finally, the relation of the NI to GNP per capita and the HDI is investigated, considering redundancy, the functional form of the relationship and outliers. For selected countries, reasons for large disparities in NI values despite similar levels of development are explored.

In Chapter 7, an estimation model is built to identify the main determinants of NI variation. The focus is on the role of overarching political, economic and social variables. Explanatory variables are chosen for their hypothesised impact on the partial indicators of the NI. The operationalisation of variables and the data sources are described in detail. The significance of relevant variables in the aggregate NI model is traced back to their influence on the NI components by disaggregation into partial indicator models.

Chapter 8 discusses the properties of the NI with reference to the desirable features of international indices set forth in Chapter 2. In addition, the NI is scrutinised in the light of widely debated objections to international indices to check if the same critique applies to this newly created measure of food security and nutrition.

Chapter 9 concludes by recapitulating briefly the steps followed in Chapters 3 and 4, and by highlighting the main empirical findings presented in Chapters 5 to 7.

Chapter 10 summarises the concept and calculation of an international Nutrition Index, and describes the empirical findings and analyses carried out with the new monitoring tool for food security and nutrition.

2 Theory of International Comparisons

According to Szilágyi (2000), international comparison theory describes a system of theoretical considerations referring to international comparisons. It is derived from methodological and analytical lessons of international comparisons carried out in the last three to four decades by international organisations, national institutes and individuals. The choice of the type of measurement and international comparison depends on the category of term that is to be measured. Szilágyi distinguishes between three different types of socio-economic categories of variables:

1. Socio-economic categories that can be quantified by a single indicator (target variable), such as household consumption, industrial production, etc. Standard methods for international comparisons of these indicators have been developed, for example purchasing power parities.

2. Categories which cannot be expressed in a scalar term according to pure theory, but conventions regarding a generally accepted indicator exist. An example is GDP as a measure of economic development, that is widely used despite its well-known deficiencies. Methods of international comparison are partly the same as for the first category of variables (PPP), and partly specifically tailored to this category (regression methods, factor analysis etc.).

3. Socio-economic categories to which no *a priori* target variable can be assigned, including subjects such as welfare, infrastructure, technical level etc. (Szilágyi 2000).

The nutrition situation clearly belongs to the last category; it cannot be fully described by a single variable or indicator, as outlined in Chapter 1.1. The focus of theoretical considerations in the present chapter is therefore on international comparisons relating to the third socio-economic category, involving the selection of a set of partial variables, their proper arrangement and combination. Prior to the steps of index formation, described in detail in Subchapter 2.3, desirable characteristics of international indicators and indices are set forth in Subchapter 2.1. Desirable properties are defined from the perspective of usefulness for the policy process as well as with respect to technical requirements. Frequent references are made to the debate on the Human Development Index, which provides great insight to the theory and practice of international comparisons and index formation.

2.1 Desirable Properties of International Indicators and Indices

A range of desirable properties has been postulated for international indicators and indices. Some of these characteristics are directly linked to the usefulness of indicators in the policy process, whereas others – though also relevant for policy purposes – are of a more technical nature and can be judged by scientific criteria.

2.1.1 Important Properties in the Policy Process

International indicators and indices should be **policy-relevant**, which means that they can assume informative, monitoring, decision-making and normative functions in the policy process. The normative impact of an indicator can be traced back to the fact that every indicator is based on a particular concept or definition with specific underlying values. If certain aspects are not properly reflected, they run the risk of being ignored in the policy debate (Nübler 1995). Ryten claims that indicators should be **adequate**, that is to say, in his words, they should 'keep on answering the right questions'. The adequacy of indicators may change in the course of the development process. Moreover, a final user and final use should be identified to assure the relevance of indicators, and they should be related to possible action; changes in the indicators should provoke questions that can serve as a basis for insights (Ryten 2000). In this sense indicators should be useful for describing living conditions, for diagnosis and analysis, and as a base for decision-making (Nübler 1995).

Ryten has presented a list of additional key attributes which a 'high profile single number' should in his opinion have. First, the number should be **communicable** in the sense that a short and simple explanation can be given to potential users to describe what the number attempts to measure. Second, it should be **replicable**, i.e. an interested user should be able to reach the same results by means of the same basic data, definitions and procedures, and sufficient resources. Third, the indicator or index should be **backed by theory** in the sense that it is grounded in a conceptual basis which can be readily explained to potential users and the general public. Fourth, the 'single number' should be **credible**, a characteristic that is based on the authority and legitimacy of the compilers, the expected use of the figure in politics, and on the perception that 'objective methods rather than partisan views'[1] have been applied to gain the figure (Ryten 2000). The requirements set forth by Ryten do indeed sound

[1] This remark may be viewed critically, because 'objectivity' in science is usually defined by a discipline's mainstream, which may be less innovative than the 'outsiders'.

sensible – be it for indicators or indices with an international or only national use. Moreover, an international indicator or index should fulfil the additional technical requirements highlighted below.

2.1.2 Technical requirements

An index – or international indicators in general – ought to show **non-redundancy** vis-à-vis indicators that are already widely used (Ryten 2000). If a new indicator does not include a significant amount of new information, there is no evident reason why it should be compiled or calculated. Redundancy is usually measured by means of the Spearman rank correlation coefficient. The problem with the judgement of redundancy, which has been hotly debated with reference to the Human Development Index, lies in the fact that no commonly accepted threshold exists. Therefore, an arbitrary threshold differentiating redundancy from non-redundancy has to be specified. In an article referring to the Human Development Index, McGillivray and White have proposed that if the rank correlation coefficient between two indicators is not significantly less than 0.90 it may be defined as 'Level 1' redundancy. If the rank correlation coefficient is not significantly less than 0.70 it is called a 'Level 2' redundancy (McGillivray and White 1993). Furthermore, analysing subsets of the total data set by means of rank correlations can provide valuable insights, because indicators which fail to pass the redundancy test for the whole data set may differ substantially in rank correlations within subsets. The gain in information by a newly introduced indicator or index is as yet only fully acknowledged if the country data set is divided into groups of equal size, e.g., quartiles according to the level of GNP per capita or the measure in question (see McGillivray and White 1993, Noorbakhsh 1998 for examples)[2].

For any indicator, **validity** is a crucial issue, meaning that the indicator is able to capture in a conceptual, as well as in a statistical, sense what it pretends to measure (Bohnet 1994). Lind (1992) defines content validity as adequacy and completeness of an indicator with reference to the relevant content of its underlying concept (he admits that this is partly a matter of judgement). Nübler (1995) claims that the indicator or index should have an adequate theoretical *and* empirical basis in order to be valid. Additionally, an international indicator or index should have the property of

[2] Furthermore, Lüchters and Menkhoff (1994) have argued that '. . . the often voiced criticism of the high correlation between GDP and HDI misses the point of the explanatory claims made on behalf of the HDI. Rather, the UNDP points explicitly to the marked differences for some countries compared with their GDP rankings.' (Lüchters and Menkhoff 1994, p.10).

comparability across countries and **over time** in order to reflect relative achievements of different countries and to monitor progress in individual countries. This statement seems obvious, but indices such as earlier versions of the Human Development Index failed to reach comparability over time, as the HDI was devised at first only for the assessment of relative achievements of different countries (this irritating characteristic of the HDI has been discussed at length in the literature, see for example Lüchters and Menkhoff 1994 and 1996, McGillivray and White 1993, Nübler 1995, Trabold-Nübler 1991 and Srinivasan 1994). Indicators may lack comparability across countries if different data definitions are used internationally, or if their meaning differs substantially according to the cultural context. These points explain the weakness of literacy as an international indicator (see Lind 1992 and Hopkins 1991). Another important property is the **ability to differentiate** countries at different levels of development. This requirement is in question if the threshold for a variable is set very low and differences in abilities or achievements with respect to the variable are not taken into account[3]. The problem is further aggravated if the indicator shows little variation for the majority of countries that are to be compared (Nübler 1995). The fact that an indicator has only small or even insignificant variation across the group of countries that are to be compared makes it unsuitable for setting up a meaningful ranking, and raises the question whether this indicator is adequate in the sense that it 'answers the right questions' for that group of countries (quoted from Ryten 2000, see also Szilágyi 2000 on insignificant differences in HDI values for industrial countries at the top of the ranking, which cannot justify large ranking differences[4]).

The requirement of **robustness** encompasses the claim that the indicator should not be too susceptible to measurement error. Furthermore, with reference to composite indices, Ryten has postulated that the aggregation function should not be sensitive to small changes in weights or functional form. The indicator or index should furthermore exhibit a certain inter-period **stability**, i.e. some constancy of characteristics over time. Too rapid shifts may incite suspiciousness, especially if the variable used for ranking is known for its slow changes. On the other hand, the indicator or index should

---

[3] Literacy may serve as an example for a low threshold here, because simple reading and writing abilities are sufficient for a person to be classified as literate. Differences in the level of reading and writing ability, which may be substantial, are not considered.

[4] Reasons for the little or negligible differences of HDI values for developed countries can be found in the fact that the components literacy and discounted GDP, making up 5/9 of the overall HDI, show little variation across these countries (Szilágyi 2000, see also Nübler 1995).

show **sensitivity** in reflecting changes fully and without delay. Flow variables may therefore be preferred to stock variables, which react slowly to policy changes, or the two may be combined[5]. In short, the indicator or index in question should be sensitive enough to provide signals if changes occur, but at the same time be time-resistant in order not to signal to excess.

Some additional issues have to be considered when it comes to calculating international indices. They are discussed below, after briefly highlighting the pros and cons of indices in the subsequent section.

2.2 The Pros and Cons of International Indices

Indices that are used for league tables and international rankings by multilateral agencies have been hotly debated after their introduction. The main arguments that have been put forward for and against composite measures such as the Human Development Index are listed in Table 2.2-1. Most of the pros and cons cited with respect to the purpose of indices apply to economic and social indicators in general. However, there is no doubt that quantitative information is essential for policy-making, though not everything can be captured in numbers - subjective matters are not readily quantifiable. The policy evaluation function of indicators in general is indeed indispensable (for examples, see the Human Development Report 2000, Chapter 5 (UNDP 2000)). Compared to single indicators, indices are likely to be more prone to a 'mystification of numbers', because the impact on the index of variations in the underlying indicators can sometimes only be seen by means of model calculations and mathematical derivations. Yet, the potential for mystification might explain part of the appeal of composite measures – a trade-off between transparency of the measure and its popularity may indeed exist. Disregard of the cultural context can be avoided if indices are not seen in isolation, but interpreted in the specific context and compared with other economic and social indicators. It should be noted that reference to cultural values has sometimes been abused as a defence against the accusation of the denial of basic human rights (UNDP 2000).

---

[5] This rather abstract requirement can be illustrated by a quote from Nübler 1995, p.174, which exemplifies the delayed reaction of an education variable in the HDI: 'The average number of years of schooling is a stock variable that relates only to the population aged more than 25 years. Hence, the HDI reflects an increase in the level of human development for the rising generation only after a lag of up to twenty years. The use of another age group or flow variables, such as the school enrolment ratio, would improve the sensitivity of the HDI and indicate current human development without a long de-

Table 2.2-1 Pros and cons of international indices

PURPOSE	
PROS	CONS
quantitative measure of progress	'illusion of precision', neglect of subjective and qualitative matters
informative value and important function as a monitoring tool	limitations because of insufficient data availability, reliability and comparability
use for the evaluation of policies	contribution to the 'mystification' of numbers
guide for decision-making and targeting	disregard of cultural context and values
TECHNICAL MATTERS	
PROS	CONS
condensation of information, facilitated use of statistics by policy-makers and the public	loss of detailed information due to aggregation
measurement of not directly observable variables, comprehensiveness	defined trade-offs between dimensions irrespective of context-specific values
reduced impact of measurement error as compared to the use of single indicators	distortions by biases in part of the underlying data
POLITICAL IMPACT	
PROS	CONS
powerful, 'eye-catching' tools for advocacy	subject to political abuse, e.g. by giving rise to unjustified complacency
setting of international goals and standards	regarded as a means to impose outside priorities on countries
ability to promote good policies and to foster a sense of competition if used in rankings	blamed for the 'intimidation' of poorly performing countries, criticism of 'naming and shaming' practices
incentives for improved data collection and national statistics	incentives for data manipulation

Source author's own presentation, inspired by Streeten 1994, Nübler 1995, Ryten 2000, Szilágyi 2000 and discussions at the IAOS Conference on Statistics, Development and Human Rights 2000 in Montreux

Indices have been frequently criticised for their technical properties. The virtue of condensing information, which facilitates the use of statistics by policy-makers and the public, is contrasted with the loss of detail due to aggregation. (Yet, in politics, the alternative to using an index is often not relying on a whole set of indicators contain-

lay.' This suggestion made by Nübler 1995 has indeed been put into practice in later versions of the HDI.

ing all the detailed information, but staring at one of the single indicators - however inappropriate it may be.)

Whereas international indices are better suited than single indicators to capture multi-faceted phenomena, weighting of the components defines trade-offs that may not be in accordance with national priorities. If measurement errors in the partial indicators of an index are independently distributed and the components are aggregated as a weighted average, the impact of measurement error decreases due to aggregation. This is exemplified in Box 2.2-1 for a hypothetical three-component index with equal weighting. However, large biases in the value of one underlying indicator for some countries may eventually translate into a distorted index value, especially if the number of index components is low.

Box 2.2-1 The role of measurement error for index calculation

The following section demonstrates in a formal way how relying on a composite index may reduce the impact of measurement error. Indeed, any indicator is likely to be measured with error. Usually, a normal distribution of errors is considered reasonable. The following explanations apply to the measurement of three indicators and the subsequent index calculation for one specific country. Assume first that there is no bias in measurement inducing systematic aberrations from the true value, i.e. the expected values of the measurement errors (μ_1, μ_2, μ_3) are zero for all indicators. Furthermore, consider the case that measurement errors in the indicators I_1, I_2 and I_3 are independent. Let all three indicators enter the index with equal weights, the sum of which is normalised to one. In a hypothetical scenario where the standard deviations of measurement errors of I_1, I_2 and I_3 are $\sigma_1 = \sigma_2 = \sigma_3 = 1$ (part A of Figure B 2.2-1-1), the standard deviation of measurement error of the resulting index, σ_{Index}, is the square root of $\frac{1}{3}$ (≈ 0.58). It should be stressed here that this reduction in standard deviation of measurement error is achieved without downscaling the index compared to the indicators. Therefore, the standard deviation of measurement error of the index divided by the absolute value of the index compares favourably to the corresponding ratio of the three indicators' sum of standard deviations of measurement error to the total of the absolute values of I_1, I_2 and I_3. In a second scenario, let $\sigma_1 = \sigma_3 = 1$ and σ_2 be increased to $\sqrt{7}$ (part B of Figure B 2.2-1-1). This value of σ_2 is the borderline case for which σ_{Index} will not exceed 1.

Box 2.2-1 (continued) The role of measurement error for index calculation

If I_3 is measured with bias as in part C of Figure B 2.2-1-1, the reduction of the standard deviation still holds given the above assumptions, but one third of the bias is conferred to the index (the formal proofs of the assertions stated here are shown in Part A 2.2-1 in the Appendix). If the assumption that measurement errors of the three indicators are independent is relaxed and the 'worst case' - perfect positive correlation – is assumed, nothing is gained in terms of reducing measurement error by aggregating the three indicators as in part A. However, there is no increase in σ_{Index}. In contrast to this, for the scenario depicted in part B, σ_{Index} increases in the case of perfect positive correlation of errors compared to the relatively lower σ_1 and σ_3, but still remains below σ_2. Perfect positive correlation of measurement errors is quite unlikely, though absence of positive correlation cannot be taken for granted. Relying on an index is in any case preferable to resorting to the most inaccurately measured single indicator. Calculating an index may in some special cases increase the impact of measurement error, but is likely to reduce it in the overall sample of countries, which is favourable for intertemporal and international comparisons.

Figure B 2.2-1-1 Distributions of independent measurement errors in three indicators and distribution of measurement errors in the index

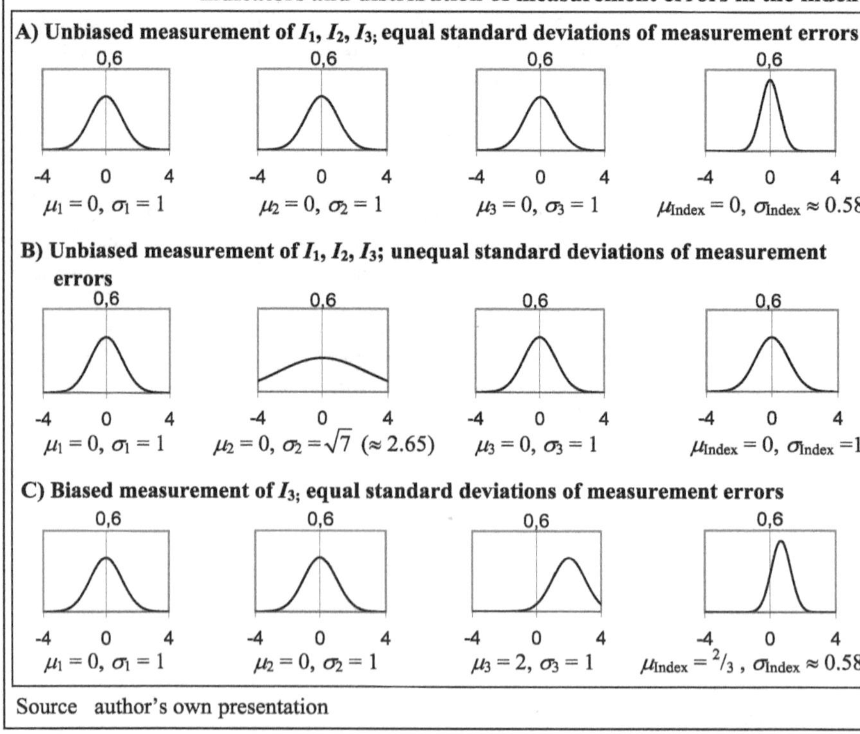

Source author's own presentation

The political impact of international indices is contentious. Both beneficial use as a 'powerful tool for advocacy' on an international and national scale and political abuse have been reported for the HDI (Handoussa 2000, UNDP 2000). The popularity of indices is a double-edged sword in this respect, since it enables the users of indices to do more good and more harm with these tools. The application of international indices and indicators for the setting of goals and for targeting has on the one hand promoted efforts of data collection and improved national statistics. On the other hand, their use as monitoring instruments and criteria for the allocation of aid has set incentives for deliberate manipulation of data. Country leaders sometimes like to make things look worse than they actually are if they seek assistance from the international community, or governments may praise their own policies by exaggerating their achievements.

2.3 A Measurement Concept of International Indices

Whilst acknowledging arguments against index calculation, the preceding discussion on the pros and cons of indices has revealed at least three favourable properties for indices vis-à-vis single indicators. First, an index composed of different, but related indicators is able to integrate several aspects of a multi-faceted phenomenon which cannot be adequately captured by a single indicator (or in other words, it can describe an 'abstract, unobservable variable' (Nübler 1995) or a 'socio-economic category to which no *a priori* target variable can be ordered' (Szilágyi 2000)). Second, the combination of indicators measured independently from each other may reduce the impact of measurement error. Third, information is condensed by means of an index, thereby enabling a quick cross country and longitudinal overview, facilitating the use of statistics especially for policy-makers and the public.

Thus, for certain purposes, indices have distinct advantages over single indicators. However, as indices themselves are based on single indicators, the concepts and steps of index formation should be considered before the calculation of the Nutrition Index. As Szilágyi (2000) states, the selection of a set of partial variables, their arrangement and combination, are a common property of methods of quantification via indices. Nübler describes an index such as the Human Development Index as a measurement concept to portray an abstract variable that is not directly observable. The measurement concept is composed of three conceptual levels:

1. an unobservable abstract variable that is not accessible to direct measurement and that is the 'ultimate criterion of interest'[6];

2. dimensions defining the abstract variable – they index the degree of success in terms of the abstract variable, but are not directly observable either;

3. empirical variables that are observable and quantifiable and that can indicate the abstract variable that is to be measured, given a theoretical relationship formulated via the dimensions (Nübler 1995).

The base of this measurement concept is the exact definition of the abstract variable and its dimensions. The interrelationship of the dimensions should be specified by a conceptual framework (Ryten 2000), and care should be taken that the concept of the abstract variable is not much broader than the included dimensions (Nübler 1995). In some cases, the concept entailed in the definition 'may be too deep and rich to be captured by any composite index or even by a detailed set of indicators' (UNDP 1999, p.127). Nevertheless, essential components should not be neglected[7] (Szilágyi 2000).

After defining the abstract variable and its dimensions, and composing a set of indicators as suggested above, various operations may have to be carried out with the indicators to obtain the index formula. The family of methods used in this process are large and manifold (Szilágyi 2000). The following sections deal with criteria for the choice of indicators, eventual transformation and standardisation of indicators, and manners of aggregation as the last step in index calculation.

2.3.1 Choice of Partial Indicators

Empirical indicators should be chosen in such a way that the dimensions of the abstract variable are captured as fully as possible. The selection of indicators for a given dimension is based on implicit assumptions and value judgements, but should be justified by theory or empirical findings. A single indicator can hardly measure a given dimension completely; validity is thought to increase when a set of indicators is used

---

[6] Examples for such abstract variables are 'human development', 'health', or 'food security and nutrition'.

[7] The HDI has been accused of not taking into account essential dimensions of human development that are entailed in the definition of human development as a 'widening of human choices', e.g., political and social freedom, respect for human rights (Nübler 1995, Desai 1991, UNDP 1990), food and nutrition (Szilágyi 2000, De Silva et al. 2000), predictability, security, upward mobility (which, in this context, means freedom from discrimination), absence of excessive inequality (Ryten 2000).

that are supplementing and controlling each other (Nübler 1995). Subsets of variables should therefore be assigned to each dimension that is identified. An optimal set of indicators is found if the size of the set is minimised and the information value is maximised (Szilágyi 2000). Or, in the words of Ryten, 'subject to constant quality, the number of components must be minimised (efficiency), . . . subject to the input being held constant, the information yielded by the number must be maximised (effectiveness)' (Ryten 2000, p.5). This can best be achieved if there is

- minimal correlation among the selected variables[8] to avoid multicollinearity, and
- maximal correlation of selected with non-selected variables – the latter requirement implies that the most characteristic indicators of a given subset are chosen.

These theoretical considerations can rarely be met in practice because, for example, the necessary information for drawing up a correlation matrix of possible variables is often lacking. The more complex the abstract variable is, the larger the set of included variables has to be. Increasing the size of the set improves the reliability of the calculations. For a small set of indicators, missing data, outliers or errors may substantially distort the final score of the respective countries (Castles 2000, cited in Szilágyi 2000). The optimal size of the indicator set may not be established due to limits of data availability, reliability and comparability, especially if a large number of countries is to be compared. 'Working with what we have' is surely a useful guideline.

As it is the combination of indicators which makes realisation of a distribution-sensitive index possible, distributional aspects of the indicators should be taken into consideration. Measures of distribution such as the Gini-coefficient of selected indicators can be chosen as additional variables, if they are available (for an attempt to build an 'Inequality-Adjusted Human Development Index', see Hicks 1997). Measures of central tendency with no natural upper boundary (such as GDP per capita) often mask inequalities and should be handled with care. Indicators based on cut-offs (e.g., various indicators of poverty and undernutrition) are less prone to obscuring inequality. Their sensitivity to distribution can even be strengthened if a variable expressing the average shortfall below a cut-off is combined with the percentage of a population falling beneath the respective cut-off, as it is done in Sen's poverty index (Sen 1976).

---

[8] According to Ryten, the components of a composite index should be independent of each other (Ryten 2000) – a requirement that is hard to fulfil due to the multicollinearity of most social and economic indicators.

Another option is the use of related indicators with cut-offs at different threshold levels[9]. Again, poor data availability may be limiting.

Furthermore the partial indicators should be chosen in such a way that the overall index meets the technical requirements set out in Chapter 2.1.2. Each single partial indicator has to be valid and comparable across countries and over time if it is not to inhibit the validity and comparability of the overall index. The selection and combination of indicators can help to bring about the characteristics of ability to differentiate, robustness, stability and sensitivity. For example, if one indicator has limited ability to differentiate between a group of countries at a medium level of development, this may be balanced by the inclusion of another indicator that differentiates within this group quite well (but might be less suited to differentiate, for example, between countries at a low level of development). Similarly, time-resistant stock variables and quickly reacting flow variables can supplement each other, assuring stability along with sufficient sensitivity of the index. *A priori* exclusion of variables with lagged effects (that is, variables which do not react immediately to a change in policy) may not be sensible, because this would discriminate against all policies pursuing long-term objectives. In the context of change over time, Ryten (2000) postulates that variations in the aggregate index should not be caused by any one of its components in particular. Special attention should be paid to the indicator in the set that is most responsive to policy (depending on the period of observation) – the component that Ryten calls the 'driver' of the synthetic index. If changes in the index are not to be dominated by one of its components, the selected indicators should not differ too much in responsiveness to policy and speed of variation. The meaning of downward movements of the selected indicators should be taken into account: if a sudden drop in one of the indicators signals a catastrophic event, whereas the other variables are subject to relatively quick recovery and may even be on the rise, the occurrence of disaster may be obscured by the aggregation of indicators (Ryten 2000). Robustness with respect to measurement error is in general expected to increase with the combination of several indicators.

Trade-offs between the fulfilment of some of these criteria exist. For example, minimal correlation among the selected variables makes the resulting index more susceptible to measurement error in one or more of the variables, and will result in

---

[9] An example of indicators that relate to the same normal distribution of a scalar variable, but refer to different cut-offs, are indicators of mild to moderate malnutrition on the one hand, and severe malnutrition on the other.

greater sensitivity to small changes in weights or functional form of the aggregation function.

After selecting the indicators, it has to be considered whether operations like transformation and standardisation have to be performed to prepare for aggregation. Available methods and their rationale are presented next.

2.3.2 Operations carried out with Partial Indicators

There are several reasons to execute mathematical or statistical operations with indicators prior to aggregation. First, some partial indicators may exhibit increasing or decreasing marginal returns to the abstract variable (or one of their dimensions, respectively), which calls for a non-linear transformation of the affected indicators. Second, there is the need to express a set of indicators with different measuring units in a common scale unit. Various methods of standardisation can be applied for that purpose.

2.3.2.1 Transformation

Concerning the transformation function of variables, decisions have to be taken on whether conversion should be performed on a linear and continuous basis (the simplest choice), or a non-linear and eventually non-continuous basis on a progressive or regressive scale. A discontinuous type of transformation is chosen if there is a change in the relation of an indicator, and the dimension it describes, beyond a certain threshold level. A regressive transformation - or, in other words, discounting of the indicator - is in place if the indicator is thought to contribute to the respective dimension of the abstract variable with declining marginal returns. Nübler (1995) has pointed out that any form of discounting should be justifiable from a theoretical or empirical point of view, e.g. by the criteria of declining marginal utility or falling social rates of return. To take an example, it has been stated that the GDP variable in the Human Development Index, which should capture the dimension 'decent standard of living' should be discounted – because there are 'diminishing returns in the conversion of income into the fulfilment of human needs' (UNDP 1990, p.13). In contrast, Rao has argued that the logarithmic transformation of the GDP variable in the first version of the HDI is inappropriate, 'since people do not compare "logs" of incomes, it is better to simply use the absolute levels' (Rao 1991, p.1455). Literacy, school enrolment rates and life expectancy, which are the indicators in the other two dimensions of the HDI (education and health), are transformed linearly. This has been criticised on the grounds of the declining marginal utility of education and falling social rates of return (Nübler 1995). Noorbakhsh (1998) has proposed a modified HDI with the components life expectancy and literacy discounted in a similar manner to GDP per capita. This snapshot of the vivid debate on the sense and non-sense of transforming particular variables of indices in a non-linear form highlights the need for some basic considerations about the non-linear transformation of partial indicators in general and discounting in particular.

In the first place, 'falling social rates of return' should be no argument for discounting of partial variables if the index is to measure human well-being – either overall well-being, e.g. the HDI, or specific well-being, e.g. with respect to food security and nutrition. If human well-being is the yardstick, declining marginal returns to human productivity should only matter indirectly. Therefore, the effect of schooling or good nutritional status on societal wealth should better be accounted for by a respective income variable than by discounting education or nutrition, both of which have a value in themselves and are direct sources of utility. The case is different for the argument about declining marginal utility. The assumption that material or immaterial 'goods' that can be accumulated by individuals are subject to diminishing marginal utility is consistent with economic theory. Examples may be income, years of schooling or food. Discounting is therefore sensible at the level of the individual. The case is more difficult if the average achievement of a group of people is to be discounted and distributional aspects are unknown. If a one unit increase in average income is brought about by income gains solely for the wealthier part of the population, decreasing marginal utility suggests that this corresponds to a lesser increase in societal welfare than if the rise originated from rising incomes of the poorest.

Discounting may hardly be justifiable for indicators expressed in percentage ratios. Discounting the literacy rate, for example, as has been done by Noorbakhsh (1998) for the Modified Human Development Index, would mean that the higher the overall level of literacy in a country the lower would be the value placed on the literacy of an individual. But to attach less importance to the achievement of literacy by the tenth person in a group than to the attainment of literacy by the first person would be at odds with equity and common sense, and would furthermore neglect the positive externalities of communication skills[10]. Similar considerations can be applied

---

[10]Discounting literacy is contradictory to the perception that in a mainly illiterate environment, inability to read and write is normally no major obstacle to participation in cultural life, communication and decision-making – simply because printed matter is not that important. In contrast, in a largely literate

to the idea of discounting the percentage of well-nourished people in a population, though the case may be a bit different. Due to the normal distribution of body indexes (height, weight, etc.), malnutrition tends to become less severe on average among the nutritionally deprived population group when the proportion of people with adequate nutritional status rises. This might be a rationale for discounting, but not a compelling one, because there are other ways to take into account the severity of malnutrition besides the application of a uniform discount rate (see some of the suggestions made in Subchapter 2.3.1 with reference to distributional aspects).

Heavy discounting is likely to be problematic if selectively applied to a specific component of an index, because it may create trade-offs that are hardly justifiable (see examples for the HDI in Ravallion 1997[11]). Furthermore, it can considerably diminish an indicator's ability to differentiate at high levels of development (Nübler 1995, Szilágyi 2000). In summary, discounting should be applied only if careful analytical considerations have shown its necessity.

2.3.2.2 Standardisation

Ryten emphasises that for the purpose of aggregation, the single indicators need to be expressed in a common metric, or need to be standardised (Ryten 2000). Szilágyi mentions the same requirement for index calculation when he says that a 'synthesis of information' is necessary that involves harmonising different measurement units (Szilágyi 2000). In the case of development indicators, it is easy to imagine that aggregating for example literacy rate and GDP per capita by simple averaging without prior standardisation would produce no meaningful results, due to the totally different measurement scales of these indicators. Similarly, with reference to nutrition indica-

---

environment, such as in a developed country, illiteracy is a severe handicap that is often experienced as painful and shameful. Therefore, raising literacy from, for example, 90 to 95% is to be valued at least as much as a rise from 30 to 35% - if not more.

[11] With respect to the 1996 version of the HDI, Ravallion has found that the following trade-offs are implied in the transformation, standardisation and aggregation of indicators: for poor countries with GDP per capita below the world mean ($5,711 per year at 1993 Purchasing Power Parity) an annual increment of $99 per capita can fully compensate for the loss of a year of life expectancy, so as to keep the HDI value constant. This seems to be a small 'price' for reduced longevity. However, the implicit monetary valuation of an extra year of life expectancy rises sharply with income. Taking the example of a country with a GDP per capita of about four times the world's mean (near Switzerland's income), its value reaches $65,038, far more than the actual GDP in the wealthiest country in the world. The construction of the HDI therefore assumes that life is far more valuable in rich than in poor countries, revealing a questionable underlying value judgement. This odd property of the HDI is to be attributed to the heavy discounting of GDP per capita above the threshold of the world's mean along with the linear transformation of life expectancy (Ravallion 1997).

tors, aggregation of unstandardised data on dietary energy supply per capita per day (ranging between 1500 and 3700 kcal per capita per day) and data on child malnutrition (for example, the percentage of underweight children with a current minimum of 0% and a maximum of 60%) would not be a sensible exercise. The case of indicators already expressed in a common metric, such as in percentage ratios, may be different. Yet, if the ranges between the maximum and minimum values of the single components differ substantially, aggregation without preceding standardisation might be questionable. Despite equal weighting, the indicator with the largest variance may dominate the overall index if no standardisation is applied, introducing a kind of 'unintentional weighting' (Noorbakhsh 1998, p.593). Szilágyi suggests three different methods of standardisation:

1. Forming the ratio of the value of country j (X_j) to the value of the country with the maximum in the given indicator.

(2.1) standardised value of $X_j = \dfrac{X_j}{X_{\max}}$

where $X_{\max}$ is the maximum value of variable X in the given data set.

2. Relating the country value X_j to estimated minimum and maximum values for variable X[12].

(2.2) standardised value of $X_j = \dfrac{X_j - X_{\min(est)}}{X_{\max(est)} - X_{\min(est)}}$

where $X_{\min(est)}$ is the estimated minimum value of variable X,
$X_{\max(est)}$ is the estimated maximum value of variable X.

3. Normalisation of X_j–values, i.e. transforming the variable so that it has a mean value of 0 and a standard deviation of 1.

(2.3) standardised value of $X_j = \dfrac{X_j - \mu_X}{\sigma_X}$

where μ_X is the mean of variable X,
σ_X is the standard deviation of variable X.

[12] Of course, it would also be possible to use the actual minimum and maximum values in the data set.

All three methods have their drawbacks. If formula (1) is applied, the standardised values of country j (and the final composite index) may not be comparable over time, as X_{max} may change over time. The same holds true for formula (2.1) if the actual minimum and maximum values are used instead of estimated boundaries. If estimated minimum and maximum values that are unlikely to be exceeded (at least in the near future) are inserted into the formula, they can only be set arbitrarily when the indicators have no a priori limits. Actual or estimated limits affect the measuring scale of the variable, which has an influence on the overall score and ranking after the combination of several standardised variables – the larger the number of variables the less the influence. At least the upper and lower limits of one and zero, set for the index by the use of formula (2.2), facilitate easy interpretation (Szilàgyi 2000). The use of formula (2.3) is likely to affect intertemporal comparability of index values also, because the mean and standard deviation of variables change over time. Taking different means and standard deviations of indicator values for normalisation (separately for the respective years) prevents comparability of index values over time. Using the mean and standard deviation of the earliest period of time for which the index is to be calculated is no real solution either, because the means and variances of the single indicators assembled in an index usually differ in pace or even direction of change. This may render the whole normalisation exercise questionable.

In conclusion it may be stated that there is no perfect way of solving the problem of expressing different indicators in a common metric. It is difficult to avoid unintentional weighting without arbitrary limit-setting whilst preserving inter-period comparability.

2.3.3 Aggregation to an Index

For the purpose of expressing a set of partial indicators in a single number, a mode of aggregation has to be defined. One simple and widely used method of aggregation is the Borda rule: N countries are ranked according to each component of the future index. The scores for the components are added up for each country, resulting in the Borda score. The Borda ranking is then obtained by ranking countries in order of their Borda score. This procedure implies equal weighting of the components. The chief advantage of the Borda rule can be seen in the fact that it only uses ordinal information, leading to purely ordinal comparisons. The obvious disadvantage of the Borda rule is related to its emphasis on ordinal distances: the ranking between China and India might alter if there is a change in one of the index components in Sub-Saharan

Africa (Dasgupta and Weale 1992). Therefore it 'violates the requirement of independence of irrelevant alternatives' set forth by Qizilbash (1997). This violation actually makes inter-period comparability impossible, and for this reason the Borda rule is not considered any further.

The alternative to a ranking rule relying on purely ordinal information is the specification of an aggregation function, including weights for the indicators. There is no unambiguous way to derive weights or choose the aggregation function. Ryten (2000) has postulated that the manner of aggregation – that is, the functional form and the weights for indicators – should be selected from an easily understandable set and should be robust in the face of small changes[13]. Nübler (1995) claims that weights should be determined by reference to a welfare function, describing the contribution of the dimensions to overall well-being. This has proved impractical so far, because such a welfare function is not known. Consequently, implicit assumptions and value judgements play an important role in the selection of weights, which should be justified theoretically or supported by empirical findings (Nübler 1995). Qizilbash (1997) has rejected the idea of arriving at universal weights for international comparisons, as this would fix trade-offs between values and contradict 'weight pluralism'. He suggests the use of a broad set of non-zero weights between zero and one (e.g., 0.1, 0.2, 0.3 etc.). If one country outranks another on all relevant weights, however assigned to the components of the index, it may be concluded that it is indeed better off. However, this implies that 'any ranking of well-being will at most be partial', because no rank order can be assigned to borderline cases (Qizilbash 1997, p.2024). The absence of a complete ranking according to a well-being index is somehow unsatisfying. However, as it is sensible to examine the sensitivity of index values to different types of weighting anyway (Ryten 2000), such an exercise may simultaneously help to check if there is room for 'weight pluralism'. If different indices calculated with different weights assigned to the partial components can be shown to be redundant vis-à-vis each other, this confirms that the index with the actually chosen weights is not too sensitive to variation in weighting. When small ranking differences are not taken too seriously, the idea of weight pluralism may find due consideration. The statement of Hopkins with

[13] Ryten has further said that 'it must be possible to claim that there are no policy applications in sight that would be disadvantaged depending on the choice of the aggregation function and its weights' (Ryten 2000, p.5).This postulate may be impossible to fulfil, though it may be stated that the mode of aggregation should not be chosen with the intent to discriminate deliberately against specific policy applications.

reference to the HDI, that it would be 'better to spend any additional resources on the quality of the underlying data rather than on interminable discussions about weights' (Hopkins 1991, p. 1471) may also be relevant for other international indices such as the Nutrition Index which is to be designed here.

As aggregation is to some extent arbitrary, the properties of the chosen aggregation function and the reasons for selecting it along with specific weights should be made explicit. In UNDP (1997), the properties of a generalised weighted average $P(\alpha)$ for aggregation of three indicators are discussed. The following version of the formula presented there is further generalised to an optional number N of partial indicators I_j weighted with w_j:

(2.4) $$P(\alpha) = (\sum_{j=1}^{N} w_j I_j^{\alpha})^{\frac{1}{\alpha}}$$

with the sum of weights normalised to one,

i.e. . $$\sum_{j=1}^{N} w_j = 1$$

The letter α denotes the order of the weighted average, with $P(1)$ being the weighted arithmetic average of the indicators I_1 to I_N. For $P(1)$ the partial indicators can be perfect substitutes of each other. For equal weighting, the effect of a marginal increase of one unit on the index value is identical for any of the indicators, irrespective of their levels. For $\alpha \to \infty$, the substitution elasticity falls to zero (proofs are listed in UNDP 1997, Technical Note 1). For $\alpha = 2$ and three equally weighted Indicators I_1, I_2, I_3 with values of, e.g., 60, 30 and 10, respectively, an increase of one unit in I_1 cannot be matched by an incremental change of one unit in I_2 or I_3. For $\alpha > 1$ and equal weighting, marginal increases in the variable with the highest relative achievement (or the largest shortfall, depending on the definition of the variable) will always dominate marginal increases in the numerical values of the other variables. Due to this important implication of the order of the weighted average, conceptual considerations on substitutability of indicators should determine the choice of the order of the aggregation function[14].

---

[14] Desai (1991) has proposed restricting the substitutability of the three basic components in the Human Development Index by using a different type of aggregation function to the simple arithmetic average, because in his opinion the capabilities that these components represent are 'co-realisable' and cannot be perfect substitutes.

Weights can be a matter of judgement, the simplest possibility being equal weighting or 'natural averaging'. Equal weighting, as well as any other assignment of weights, should be justified by theoretical or normative considerations. If quite strong correlations among the chosen variables exist[15], principal components analysis as a special form of factor analysis can be a useful tool for deriving empirically founded weights (principal components analysis is not suitable for loosely correlated variables). The general purpose of factor analysis is the reduction of a large set of multivariate information to a smaller set composed of one or more composite unobservable factors (Agung 1996). Therefore, it is a technique that serves the condensation of information, which has been cited as one of the virtues of index formation. If the dimensions of the abstract variable are represented by sets of variables, with a varying number of variables in each set, it may be desirable to externally introduce weights into principal components analysis. This is useful as it prevents those dimensions represented by more variables from gaining relatively more influence on the resulting index via the sum of the weights of their variables. Adjustments can be achieved by a weighted principal components analysis (De Silva et al. 2000). The implementation of the methodology of principal components analysis for the determination of weights is described in Chapter 4.3.

[15] Despite the claim that there should be 'minimal correlation of selected variables' (see Section 2.3.1), the correlations among the chosen indicators may be quite strong owing to unavoidable multi-collinearity of social and economic indicators.

3 The Concept of the Nutrition Index (NI)

The purpose of the Nutrition Index (NI) is the quantitative description of a country's or region's food and nutrition security. The participants at the World Food Summit in 1996 agreed upon the following definition of food security: 'Food Security exists when all people, at all times, have physical and economic access to sufficient, safe and nutritious food to meet their dietary needs and food preferences for an active and healthy life' (FAO 1996a, Paragraph 1). The definition of nutrition security goes even further: 'A person is considered nutrition secure when she or he has a nutritionally adequate diet and the food consumed is biologically utilised such that adequate performance is maintained in growth, resisting or recovering from disease, pregnancy, lactation and physical work' (Frankenberger et al. 1997, p.1). Besides food security, adequate care and health environments are needed to achieve nutrition security.

It is evident from both definitions that food and nutrition security are complex phenomena and that the criteria set forth cannot be captured by any single indicator. The food and nutrition security situation in a country – occasionally referred to as 'nutrition situation' in this text - may be regarded as an abstract variable that is not accessible to direct measurement. As has been outlined in the preceding chapter, the dimensions of this abstract variable have to be defined prior to the selection of empirical variables for the Nutrition Index. For the purpose of index design, an outcome-oriented definition of dimensions is sought that is suitable for practical purposes. This definition is based on the conceptual framework presented in 3.1, before the selection of indicators is discussed in Subchapter 3.2.

3.1 Conceptual Framework

The conceptual framework for the nutrition situation to be measured by the NI is shown in Figure 3.1-1. A distinction is made between three levels of causality corresponding to basic, underlying and immediate causes of hunger and malnutrition, and malnutrition-related deaths. The focus is on immediate causes and effects at the individual level and the outcomes for individuals, though the pathways leading from basic to underlying and from underlying to immediate causes will be briefly described, as they are relevant for modelling the NI in Chapter 7.

Figure 3.1-1 Determinants, effects and outcomes of hunger and malnutrition

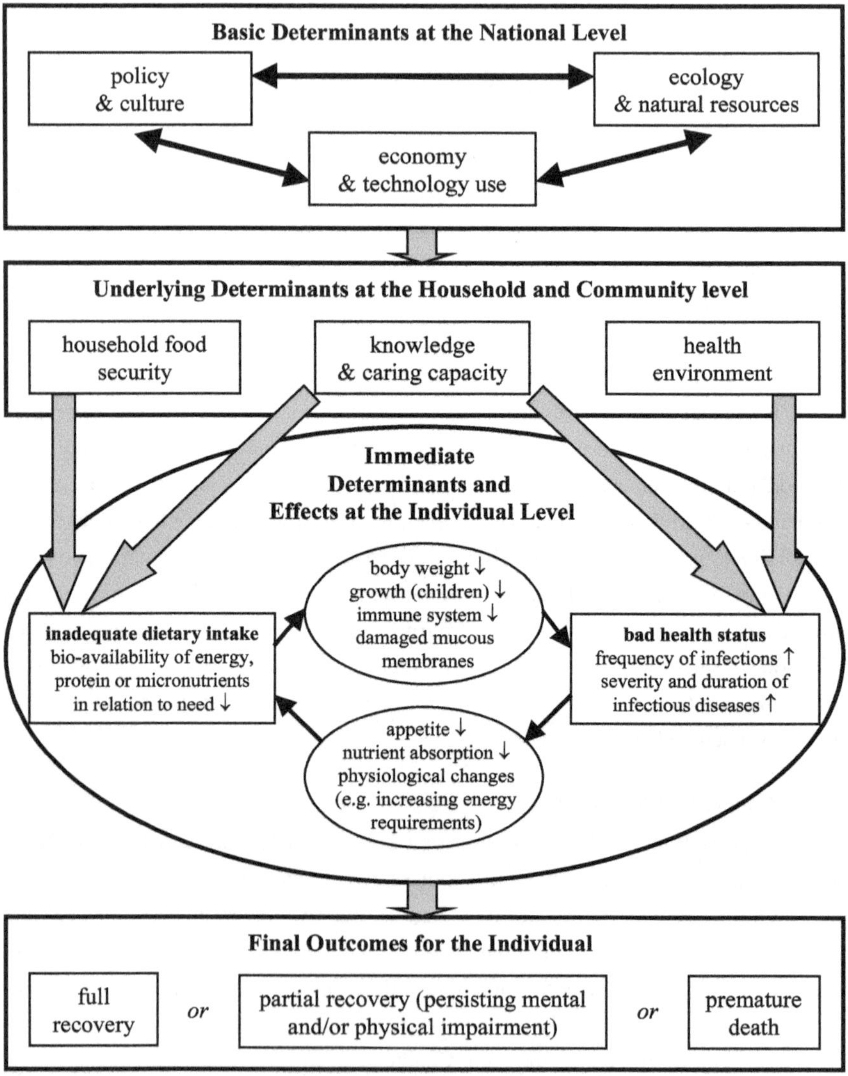

Source adapted from UNICEF 1990b, Smith and Haddad 2000, von Braun et al. 1998, Tomkins and Watson 1989

The basic determinants of hunger and malnutrition at the national level relate to the interacting fields of economy and technology, policy and culture, ecology and natural resources: scarcity of natural resources, lack of technology and economic assets, market and policy failure (including violent political conflicts and discrimination) lead to low macro-economic performance. The result is widespread poverty, especially in the absence of mitigating social policies. Poverty seriously interferes with household food security and caring capacity[1] in the family and in the community, and impedes investment in education and utilisation of health resources. Moreover, the underlying determinants are directly linked to human rights, not only depending on the right to food, medical care and other basic necessities, but also on the right to social security and protection, equal access to public services, special care and assistance for motherhood and childhood, and on the right to education and gender equity. Whenever these rights are violated, be it due to vast urban-rural gaps in public service provision, the absence of social security for the poorest, or heavy discrimination against women (who are in general the main care-takers of children), the risk of hunger and malnutrition is likely to increase. Furthermore, household food security is directly affected by the functioning of local markets, the use of technology in households (e.g. for food conservation), agro-ecological factors such as soil fertility and climate, and by policy measures such as food for work programmes or food subsidies. The acquisition of knowledge is influenced by education policy and public health campaigns. Women's education and their caring capacity are frequently hampered by culturally embedded and politically reinforced discrimination. Time resources as another important factor for care-giving are determined by the workload and division of labour, available technologies (for example, firewood-saving stoves), natural resources (e.g. water) and cultural habits. Health environments are shaped by the safety of workplaces, natural conditions and policy outcomes like public investments in the provision of water, sanitation, health care and public health programmes such as immunisation campaigns (von Braun et al. 1998, Smith and Haddad 2000).

The impact of the underlying determinants (household food security, knowledge and caring capacity, and health environments) on the immediate causes of dietary intake and health status is obvious: what kind of food, and how much, is accessible to an individual household member is determined by the availability of food in the house-

[1] The term 'caring capacity' is defined here as the amount of time, personal effort and material resources that can be mobilised for care-giving (adapted from Smith and Haddad 2000).

hold, by the recognition of personal needs and the needs of others, and by the caring capacity of the family members responsible for food preparation and the feeding of children. Knowledge, time, and material resources other than food are necessary to care for oneself and others by preventing and curing infectious diseases (UNICEF 1990b). The exposure to pathogens and therefore probability of illness is partly a function of the health environment. Depending on the living conditions as determined by food security, knowledge, caring capacity and the health environment, the individual can get trapped in a vicious circle of inadequate dietary intake and bad health status. The starting point can be either energy, protein, and/or micronutrient intake falling below requirements. The vicious circle may also get started by an infectious disease - therefore bad health environments and the lack of knowledge or caring capacity can partly offset the positive effect of sufficient food availability on nutritional status. The consequences of lack of dietary energy and protein and of the most prevalent micronutrient deficiencies[2] are shown in Table 3.1-1, with special reference to the effects mentioned in Figure 3.1-1. The consequence of acute lack of dietary energy intake is reduced body mass (wasting). Moreover, chronic lack of dietary energy, protein and some micronutrient deficiencies cause growth faltering in children (stunting). Both wasting and stunting are reflected in low body weight with reference to a child's age (underweight). Other effects of dietary energy, protein and/or some micronutrient deficiencies are reduced functioning of the immune system and damage to mucous membranes. Barriers against infection are no longer intact, and resistance to pathogens is weakened. The risk of viral and bacterial infections rises, and a bad nutritional status prior to infection prolongs the duration and aggravates the severity of the infectious disease. Loss of appetite is a frequent consequence of infection, which in turn reduces dietary intake. Physiological changes specific to many infectious diseases are energy-consuming fever and catabolism of body protein that enhance the need for dietary energy and protein (Tomkins and Watson 1989).

The final outcome for the human being depends on genuine individual factors, care, health and nutrition interventions from the outside, and on the intensity of adverse effects in the vicious circle. At best, the individual gains full recovery after some time of suffering. Partial recovery is another possible outcome; in this case, reduced body size in adulthood and irreversible damage to mental capacity, blindness or other

---

[2] Any of the listed nutrient deficiencies can be the single cause of the effects noted in Table 3.1-1, though in reality multiple deficiencies, with often fatal synergies, are most prevalent.

disabilities are a remaining characteristic of the person affected[3]. Premature death is the most serious outcome of the vicious circle (Pinstrup-Andersen et al. 1995).

Table 3.1-1 Effects of the most prevalent macro- and micronutrient deficiencies

Deficient Dietary Component	**Physiological Changes and Clinical Symptoms**				
	Weight Loss 'wasting'	**Growth Faltering in Children 'stunting'**	**Reduced Immune Resistance**	**Damage in Mucous Membranes**	**Specific Symptoms**
Macronutrients					
Energy[1]	X	X	X		Muscular Atrophy
Protein[1]		X	X		Oedema
Micronutrients					
Vitamin A		X	X	X	Bitot's Spots, Blindness
Iron			X	X	Anaemia
Iodine[2]		(X)			Goitre, Cretinism

Source Biesalski 1995, Spittler et al. 1995, Barth et al. 1995, de Gruyter 1998

Notes 1 mostly energy and protein deficiency are combined, as the so-called Protein-Energy-Malnutrition (PEM)

2 only the most severe form of iodine deficiency leads to growth faltering, which is associated with cretinism

Regarding the characterisation of a country's nutrition situation, it might be tempting to define it via the final outcomes for individuals depicted in Figure 3.1-1. However, in this case, the suffering that already starts with food insecurity at the household level and having to worry about the daily meals would only be considered if it resulted in permanent disabilities or death. Similarly, the human misery associated with the vicious circle of inadequate dietary intake and bad health status would not be taken into account if the individual gains full recovery. On the other hand, an operational definition solely based on adequate food supplies at the household level would miss impor-

[3] Of course, not only well-being, but also working and learning capacity are strongly affected by these impairments in present and later life. The formation of human capital, a driving force of economic growth, is in the long run impeded by the effects of malnutrition on human health. Therefore, another pathway leads from the vicious circle and its adverse final outcomes back to the 'economy' as a basic determinant, which is not depicted in Figure 3.1-1.

tant aspects, too. As is evident from the above, the availability of sufficient and nutritious food at the household level does not guarantee that food intake meets the dietary requirements of individual household members, nor does it imply that health status permits the biological utilisation of food.

Therefore, the selected definition goes beyond mere food availability, considering the outcomes of insufficient quantity, quality or safety of food as well as the consequences of a failure to utilise nutrients biologically. The following criteria are proposed for the judgement of a country's nutrition situation: The vast majority of the population, especially vulnerable groups like children:

- are free from hunger (have adequate dietary energy supply);
- have good nutritional status (as measured by anthropometrics);
- and are safe from death caused directly or indirectly by undernutrition.

Clearly, the nutritional well-being of the individual is at the centre of this definition[4]. Figure 3.1-1 gives a basic understanding of the interrelationship of the above dimensions, as it has been put forward in Chapter 2.3. Freedom from hunger is closely related to household food security and refers to dietary intake as one determinant of nutritional status. Since the NI is meant to capture the problems of hunger and undernutrition – including micronutrient deficiencies - but not overnutrition and related chronic diseases, a 'good nutritional status' is solely defined by the absence of anthropometric shortfalls. Good nutritional status is essential to minimise the risk of premature death and cannot be achieved if health status is inadequate or if protein and vital micronutrients are lacking (compare Table 3.1-1). These criteria for the assessment of the nutrition situation form the base for the choice of indicators that is discussed next.

3.2 Choice of Indicators

The choice of indicators is guided by the above definition of a good nutrition situation. As far as possible, it is carried out in accordance with the criteria for the selection of variables stated in Chapter 2.3.1. The following indicators are chosen to reflect the three dimensions of the nutrition situation:

- the percentage of undernourished as estimated by FAO, reflecting the share of the population with insufficient dietary energy intake,

[4] This definition is based on objective criteria and does not entail subjective aspects of food and nutrition security. In contrast, subjective aspects can be found in the FAO definition of food security that includes access to food to meet not only dietary needs, but also food preferences (FAO 1996a).

- the percentage of underweight children under the age of five, indicating the proportion of children suffering from weight loss and/or reduced growth,
- the under-five mortality rate, reflecting to a large extent the fatal consequence of the synergy between inadequate food intake and bad health environments.

It might be desirable to assign more than one indicator to each dimension, but deficiencies in data availability are limiting. A comparative characterisation of the indicators is given in Table 3.2-1 (for the statistical definitions, see Chapter 4.1). It is obvious that they refer to different, but complementary aspects of a bad nutrition situation. A common feature is that undernourishment, underweight in children and child mortality are associated with micronutrient deficiencies. Even though no indicator for vitamin or mineral deficiencies was included in the index due to insufficient data availability, the NI is likely to reflect the lack of micronutrients to a certain extent (see Chapter 6.1.2 for an empirical analysis).

In the absence of precise international data on the food consumption of households and individuals, as an indicator the percentage of undernourished is preferred to dietary energy supply per capita (DES) to reflect the dimension of adequate dietary energy intake. Though undernourishment as estimated by FAO may not give due consideration to inequality of dietary energy intakes (Svedberg 1998 and Smith 1998), dietary energy supply per capita is even less likely to do so[5].

Moreover, there is no fixed upper boundary of dietary energy supply per capita that guarantees freedom from hunger for the total population – for this reason, arbitrary limit setting would be unavoidable if dietary energy supply per capita was used for index calculation (see Chapter 2.3.2.2 for a discussion of related problems).

---

[5] Dietary energy supply per capita is a measure of central tendency like GNP per capita. As has been argued in Chapter 2.3.1, measures of central tendency are in general less sensitive to distributional aspects than cut-off based indicators (the percentage of undernourished belonging to the latter category). In contrast to incomes, dietary intakes have certain upper limits given by physiological boundaries, and for this reason their distribution is less unequal than income distribution. However, overconsumption of dietary energy by the majority of the population may still mask the food deprivation of a poor minority. At high levels of dietary energy supply per capita and low prevalence of undernourishment, a significant increase in dietary energy supply per capita may arise from more excessive intakes of the overnourished majority as well as from a reduction in the food deficit of the undernourished.

Table 3.2-1 Comparison of common indicators for the nutrition situation

Criteria	Single Indicators for the Nutrition Situation		
	Percentage of Undernourished	**Percentage of Underweight Children**	**Under-Five Mortality Rate**
Aspect of Nutritional Problem	inadequate dietary *energy* intake ('hunger')	low body weight with reference to height and/or age ('malnutrition')	death in childhood ('mortality associated with malnutrition')
	immediate cause of the vicious circle[1] and possible effect of infection	cause and effect within, eventually final outcome of vicious circle[1]	possible final outcome of vicious circle[1]
	important, but narrow aspect of nutritional well-being indicated; not congruent with, but frequently associated with inadequacy of micronutrient intake	broad aspects of nutritional well-being and nutritional status, outcomes of dietary energy, protein and micronutrient deficiencies and unhealthy living conditions captured	central aspect of well-being described, to a large extent outcome of dietary energy intake, some micronutrient deficiencies and infection
	interaction of individual constitution, food intake, care and health environments not considered	interaction of individual and external factors (food, care, health environments) reflected	interplay of individual resilience and external factors reflected, but causes of death other than starvation or its potentiating effects on mortality included
Reference Group	whole population	population subgroup: children under 5	population subgroup: children under 5
	comprehensive, but no consideration of different vulnerability of population subgroups	especially vulnerable group; serious and irreversible consequences of malnutrition on well-being, learning and working capacity	especially vulnerable group, central role of nutrition for survival

Source Tomkins and Watson 1989, Svedberg 1998, UNICEF 1998, Biesalski 1995

Note 1 vicious circle of inadequate dietary intake and bad health, see Figure 3.1-1.

Table 3.2-1 (continued) Comparison of common indicators for the nutrition situation

Criteria	Single Indicators for the Nutrition Situation		
	Percentage of Undernourished	**Percentage of Underweight Children**	**Under-Five Mortality Rate**
Consideration of Inequality	inequality of inter-household food allocation within the population considered	inequality of allocation of food and other resources primarily within the subgroup of children under five	
	no information about inequitable intra-household allocation included	hints on inter- and intra-household allocation of resources within the population	
Origin of Data	estimates derived mainly from macro-data on distribution, production, trade and use of agricultural products	nationally representative surveys as data sources, supplemented by estimates	government statistics, representative surveys and partly estimates

Source Tomkins and Watson 1989, Svedberg 1998, UNICEF 1998, Biesalski 1995

The combination of the percentage of undernourished, that refers to the population as a whole, and the two indicators relating to children under five has advantages with regard to the resulting index: both the nutrition situation of the whole population and the effects of inadequate nutrition (and concurrent factors) on a physiologically and socially very vulnerable group are captured by the NI. Besides the fact that children's nutritional status deserves particular attention because malnutrition puts them at high risk of physical and mental impairment and death, relying on data about children is also justified from a methodological point of view: anthropometric data on children are favourable due to their international comparability. In contrast to the body size adults can reach, the growth potential of children under five most probably does not differ significantly by ethnic origin (Svedberg 1998). Moreover, international data about the nutritional status of adults are not yet broadly available, whereas anthropometric data referring to children have been collected by WHO in a large, regularly updated database (WHO 2000a), so that data availability suggests their use.

Three anthropometric indicators for children are usually collected in nutrition surveys: the prevalence of low height for age (that is stunting), low weight for height (the so-called wasting), and low weight for age (i.e., underweight). Wasting is not well suited for use as a single anthropometric indicator in the Nutrition Index. The prevalence of wasting tends to be much lower than the prevalence of stunting and underweight, which means that the weight for height indicator is not able to capture the true magnitude of the problem of child malnutrition – it is not „adequate" as required in Chapter 2.3.1. Concerning stunting in children, it basically follows the same trends as underweight (WHO 1997). However, growth faltering does not occur immediately if food and nutrition security suddenly worsens, but with a certain time lag. Therefore, the ability of this indicator to reflect 'changes fully and without delay' (see Chapter 2.1.2) is limited, because it does not indicate acute malnutrition. The anthropometric measure for underweight – low weight for age – is preferred to the other anthropometric indices because it encompasses both chronic and acute malnutrition. Furthermore, among the subset of the three anthropometric measures, underweight is the most characteristic indicator, since it has the highest correlations with the other two indicators – the prevalence rates of stunting and wasting are only relatively weakly correlated (correlation coefficients were computed for data released in WHO 1997)[6].

A shortcoming of anthropometric indicators can be seen in the fact that they do not reveal the most tragic consequence of undernutrition: premature death. Equal levels of malnutrition in a country as measured by low weight for age can have quite different effects on the proportion of malnutrition-related deaths among children, depending on the overall level of child mortality (compare the example of Niger and Yemen described in Box 3.2-1 on page 42). This disadvantage of the percentage of underweight children is balanced by the inclusion of the under-five mortality rate. Clearly, the mortality data comprise other causes of death than malnutrition, and the actual contribution of malnutrition to mortality is not easy to track because the proximate cause of death is frequently an infectious disease (Pelletier et al. 1994). However, about 54% of deaths among children under five in developing countries are estimated to be associated with malnutrition (Bailey et al. 1998). Mortality data are also available for the population as a whole, but mortality among adults varies to a larger extent according to factors that are not linked with nutrition (like lifestyles, hazardous occupations and active participation in wars). Therefore, the under-five mortality rate is preferred as a proxy variable for premature deaths associated with malnutrition.

---

[6] An alternative to the inclusion of the prevalence of underweight would be the use of all three anthropometric indicators or of stunting and wasting rates only. This, however, would put higher requirements on data availability.

Another type of complementarity between the indicators selected for the NI exists with respect to the kind of inequality in resource allocation[7]. Differences between the indicators in the reflection of inter-household and intra-household allocation of food and other supplies are mainly rooted in the reference groups used and in the ways the data are derived. The primary source for the indicators referring to children are micro-data gained in representative surveys. FAO approaches the problem of quantifying undernourishment from the other side: macro-data like agricultural trade and production statistics, and data or estimates about the national distribution of consumption are the point of departure. Concerns have been raised about the validity of these underlying data and the resulting estimates (Svedberg 1998). The fact that the percentage of undernourished is basically an inverse, regressive transformation of dietary energy supply per capita (see the model presented in Chapter 4.1.1) indeed suggests that the inequality of dietary energy intakes may not have found due consideration in the estimation process done by FAO[8].

In any case, errors in the underlying data are more likely to affect the absolute number of people estimated to be undernourished than the relative differences in the proportion of undernourished among countries or in succeeding periods of time; the latter are far more relevant for index calculation. On the other hand, the validity of the data about children is restricted by sampling and estimation errors and possible small inadequacies in international reference standards (compare Klasen 1999) and, in the case of the under-five mortality rate, partly dependent on the reliability of government statistics. Concerning data quality, there is certainly a lot of scope for improvement –

---

[7] The consideration of inequality is of course limited in the undernourishment and underweight indicators, because undernourishment and underweight can vary gradually among individuals from severe to mild, but the data are derived by using cut-off points.

[8] However, whereas the dominant role of DES for the estimation of undernourishment (compare Table 4.1.1-1 in Chapter 4) raises doubts about the validity of the FAO data, the inverse, regressive transformation itself (see Figure 4.1.1-2) is very plausible. A linear transformation of DES would award the same credits to a given increase in dietary energy supply per capita, irrespective of the initial level of DES. In contrast, the regressive transformation – i.e., the discounting - implies declining marginal returns of rises in DES to the reduction of the percentage of undernourished, which means that increases in dietary energy supply per capita tend to decrease the percentage of undernourished the less the higher the initial level of DES. For example, a rise in DES from 2000 to 2500 kcal is expected to reduce the proportion of undernourished in the population by about 25 percentage points, whereas an increase in DES from 3000 to 3500 kcal would result in a reduction by about only two percentage points according to the model presented in Chapter 4.1.1. Assuming diminishing returns of rises in dietary energy supply per capita to food security and nutrition is in line with the arguments put forward in Chapter 2.3.2.1, and also with empirical findings on the determinants of child malnutrition (Smith and Haddad 2000).

but the assessment of the nutrition situation should not be postponed until perfect data with global coverage are available[9].

As all three indicators selected are expressed as share of a total, they have *a priori* limits (0 and 100%). This is a potential advantage for index calculation, because it may render standardisation methods prior to aggregation, which entail several drawbacks (compare Chapter 2.3.2.2 for an overview), completely unnecessary. For the reasons discussed at length in Chapter 2.3.2.1, a regressive or progressive transformation of partial indicators is not considered adequate for indicators of food security and nutrition defined as headcount ratios[10]. Thus, possible difficulties that might be brought about by a non-linear transformation of part of the index components can be avoided for the chosen set of indicators.

For the aggregation of the three indicators, weighting factors have to be determined. Factor analysis was considered appropriate for the derivation of weights, because the conceptual relationships of the indicators suggest the existence of strong correlations, and because this technique allows the use of weighting factors that are based on empirical evidence. The details concerning the aggregation of indicators are discussed in the next chapter together with other methodological issues, like data refinement and estimation procedures.

Box 3.2-1 Malnutrition and related deaths in children in Niger and Yemen

Anthropometric data do not tell the whole story about the tragedy of malnutrition. This is exemplified here for Niger and Yemen, where quite similar prevalences of underweight in children under five were found with 42.6% (in 1992) and 46.1% (in 1997), respectively. Yet, figures on the percentage of underweight children alone do not reveal the proportion of children dying due to the potentiating effect of malnutrition on disease, because this share depends on the overall mortality level. But the *relative contribution* of malnutrition to all causes of death varies independent of mortality levels according to the prevalence and severity of the malnutrition. Based on this finding, the share of child deaths caused by the potentiating effect of malnutrition on disease can be roughly estimated with the prevalence of malnutrition as a single input (Pelletier et al. 1994).

---

[9] Data availability and comparability problems are addressed in more detail in the next chapter.

[10] As already mentioned, the current practice of calculating the percentage of undernourished may be regarded as a plausible way of discounting dietary energy supply per capita, see Chapter 4.1.1.

Box 3.2-1 (continued) Malnutrition and related deaths in children in Niger and Yemen

This methodology was applied to the anthropometric data from Niger and Yemen, and malnutrition could be estimated to contribute by about 55% and 58% to child mortality. Taking into account the large divergence of child mortality levels in both countries – in Niger, 320 out of 1000 children die before their fifth birthday, but only 100 per 1000 in Yemen - the implications are quite different. Though the follow-up of a cohort of children five years after birth would reveal that about 44% are alive and above the underweight cut-off in both countries, the fate of the rest is not the same. Whilst in Yemen children's lives and health have been adversely affected by malnutrition in 52% of the initial cohort nearly five years later, the respective number is 42% in Niger. But in Niger, malnutrition has brought death to more than two fifths of this group, whereas this happened to only one ninth of these children in Yemen. In Niger, about 18 out of a cohort of 100 children are no longer alive five years after birth due to the potentiating effect of malnutrition on disease. In Yemen, only 6 out of 100 children have died from this cause. Therefore, in Niger, three times as many child fatalities are attributable to malnutrition than in Yemen (see Figure B 3.2-1-1). Though the anthropometric data suggest that the problem of malnutrition is slightly worse in Yemen, a look at malnutrition-related deaths does not confirm this notion. This provides strong arguments for integrating child mortality data in the Nutrition Index.

Figure B 3.2-1-1 Shares of a cohort of children affected by malnutrition, malnutrition-related deaths, and deaths from other causes five years after birth

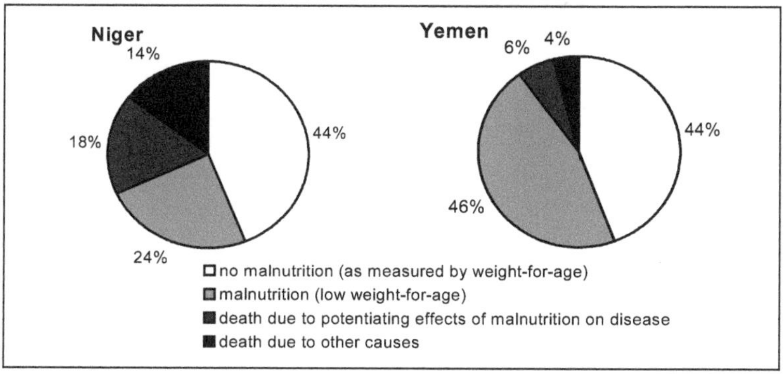

Source own presentation based on calculations with data from Kourgueni et al. 1992, Central Statistical Organisation, Sana'a,Yemen 1998, UNICEF 1995 and UNICEF 1999

Note Evidently, the percentage of underweight children is lower when it is referred to the whole cohort of children born five years ago instead of to the surviving children only, as is done in the nutrition surveys. Moreover, underweight prevalence reported for children at age four - taken as a proxy for underweight prevalence in children just before their fifth birthday - is lower than prevalence in all under-five-year olds in Niger and slightly higher in Yemen.

4 Underlying Data and Methodology

Insufficient data availability and comparability limit the usefulness of international indicators and the scope for index calculation, as has been mentioned in Chapter 2.2 (see Table 2.2-1). Fortunately, international nutrition monitoring systems have improved over the past two decades, and the published data now permit presentation and analyses which were not possible in the 1980s. Still, some data transformations and estimates are necessary for some countries, and some variables, in order to compile a comprehensive data set. Faced by a lack of direct nutrition-related information for some countries, ancillary criteria are defined in order to identify countries that are main candidates for nutritional surveillance. Thereafter, data gaps are supplemented as far as possible by the author's own estimates, and estimates from other sources. Refinement procedures are applied to the anthropometric data, and reference periods are fixed for the three partial indicators. These are necessary steps prior to the derivation of weights in factor analysis and the subsequent aggregation of NI components in Chapter 4.3, since factor analysis presupposes empirical data. The sensitivity of the NI to modifications of the aggregation function - arising from different ways of standardising the partial indicators and variations of the weights - is tested in an analysis of redundancy of several versions of the NI. The methodologies used are outlined as follows.

4.1 Data Availability and Comparability

The calculation of the NI is not possible for every country and for any point in time due to lack of data for some countries and years. Whereas annual data for the under-five mortality rate of nearly all countries in the world are generally available, data availability for the percentage of undernourished and the proportion of underweight children is quite limited. For industrialised countries, virtually no data about undernutrition exist. However, nutritional surveillance is not a priority in most developed countries, because wealth and well-established social safety nets have led to a good nutrition situation for the population overall. The efforts of data collection and monitoring should focus on countries where considerable proportions of the population are still suffering from hunger and are physically and mentally affected by malnutrition, and these countries are the main candidates for the following estimation procedures designed to close data gaps.

Thus, a first step undertaken in the construction of the index is the division of countries into those with problems of undernutrition and those without. For countries with a lack of anthropometric data, the under-five mortality rate and dietary energy supply per capita can serve as auxiliary criteria. A look at available data for dietary energy supply per capita (mean 1995-97) and undernourishment (1995-97) reveals that in the range of more than 3,100 kcal per capita per day, less than 2.5% of the population are considered to be undernourished in most countries, that is, they fall into the 'very low hunger category' according to FAO (FAO 1999a)[1]. But taken as a single auxiliary criterion, the national average of dietary energy supply might conceal disparities in inter- and intra-household food allocation. Moreover, as has been outlined in the theoretical part, there is more to adequate nutrition than sufficient dietary energy intake. These facts are illustrated by the situation in some North African and Near Eastern countries - Egypt, Turkey and Syria - where dietary energy supply per capita even exceeds the mark of 3,100 kcal, whilst more than 10% of children under five are undernourished. At the same time, the under-five mortality rate in these three countries amounts to more than 30 per 1,000 and is therefore considerably higher than in countries that have equal or even lower dietary energy supply per capita, but have reached a higher stage of economic development. In Japan, for example, the average citizen consumes less than 3,000 kcal per day, but the under-five mortality rate is as low as 6 per thousand live births, and there is no reason to assume malnutrition problems in this highly industrialised country[2]. Therefore, the under-five mortality rate and dietary energy supply per capita are taken together for a rough assessment of the severity of nutrition problems. A country is assigned to the group with no severe nutrition problems if dietary energy supply per capita exceeds 2,900 kcal per capita per day *and* the under-five mortality rate is below 15 per thousand. These criteria are quite stringent in order to make sure that no country with persisting undernutrition escapes nutritional surveillance[3]. Of course, there are borderline cases, and it is accepted that some coun-

[1] Compare the model laid out in Chapter 4.1.1: the predicted value for dietary energy supply per capita of 3,100 kcal is 2.5% undernourished.

[2] The latest nutrition survey in Japan was conducted in 1980, and the percentage of underweight children under five was no more than 3.1%.

[3] Malaysia, Slovakia and Kuwait are odd cases that are included in the group with nutrition problems, though they meet the criteria: in Malaysia, underweight prevalence in children was nearly 20% in 1995; in Kuwait, the percentage of undernourished was reported to be 3% for 1995-97, but for the preceding period 1990-92, when the war against Iraq took place, 27% undernourished was estimated; dietary energy supply per capita in Slovakia was only slightly higher than 2,900 kcal in 1995-97 whilst the under-five mortality rate (11 per 1000) exceeded the level in Western industrialised countries.

tries lacking data, where nutrition problems might actually be smaller than had been assumed, may be erroneously classified as 'countries with severe nutrition problems'.

Estimating the percentage of undernourished for countries with probably severe nutrition problems, but lacking data, is the next step.

4.1.1 Percentage of Undernourished

The FAO method of estimating the percentage of undernourished is based on three parameters: dietary energy supply per capita, the coefficient of variation of dietary energy intakes across households, and minimum dietary energy requirements. Dietary energy supply per capita, which is based on national food availability (derived from macro-data on agricultural production, net trade flows and stock changes, as well as uses other than food consumption in the Food Balance Sheets) (FAO 2000b), is taken as a proxy for dietary energy intake. The coefficient of variation of dietary energy intakes is obtained from national food consumption surveys or calculated from household income and expenditure data. (The latter method is chosen if there are no national food consumption surveys that would allow to derive such coefficients of variation.) The cut-off point of minimum dietary energy requirement is computed for the 'average' inhabitant of a country, taking into account the composition of the population by age and sex and respective dietary energy requirement norms (FAO 1996b). The percentage of undernourished is then derived from the approximated distribution of dietary energy intakes (which is assumed to be log-normal and is based on dietary energy supply per capita and the coefficient of variation of dietary energy intakes) and the minimum dietary energy requirement cut-off. Concerns have been raised about the reliability of all three underlying parameters and the resulting estimates of undernourishment, and the fact that the FAO methodology does not consider intra-household food allocation has been criticised[4]. Though FAO seeks to take into account the variation in dietary energy intake across households, it is unlikely that the results are satisfactory in this respect due to the paucity of data on distribution (Svedberg 1998). However, more convincing methodological alternatives to estimate undernourishment on an international scale have not yet been developed. As has been argued in Chapter 3.2, the percentage of undernourished assessed by FAO can be expected to be more

---

[4] These criticisms provide a strong rationale for building a composite measure like the NI with the intention of both limiting the impact of measurement errors of single indicators, and of combining the food-supply-oriented approach with data reflecting the intra-household allocation of food, such as anthropometric data and child mortality data.

sensitive to inequality than dietary energy supply per capita (see also Chapter 6.2.1, where the NI, the percentage of undernourished and dietary energy supply per capita are tested for their sensitivity to income distribution against the yardstick of GNP per capita).

The estimates of undernourishment used for the Nutrition Index are based on three-year periods and have been published for 1979-81, 1990-92 and 1995-97 for developing countries with more than one million inhabitants (FAO 1999a). In spite of rising concern for the nutrition situation in parts of Eastern Europe and in the countries of the former Soviet Union, no country estimates of undernourishment were published by FAO before 2000[5]. For the Eastern European and Central Asian region and a few developing countries with less than one million inhabitants, estimates were therefore generated by a simple model to complement the data set. If information about dietary energy supply per capita, the distribution of dietary intakes according to actual consumption or income data, and the average minimum dietary energy requirement was available for every country, the data set needed for the calculation of the NI could be completed following the methodology used and published by FAO. In the absence of some of these data, linear regression (ordinary least squares, OLS) was chosen as the technique to reproduce the FAO figures as closely as possible[6]. Regressions were run with combinations of dietary energy supply per capita (1995-97), its logarithm and square as independent variables, and the percentage of undernourished (1995-97) as a dependent variable[7]. The best fit, indicated by the highest adjusted R-square, was yielded by using a linear combination of dietary energy supply per capita (DES) and its logarithm only. As a next step, regressions were carried out with dietary energy supply per capita and its logarithm in the same way, but for different periods of time -

---

[5] Estimates of undernourishment for this region were released in October 2000 and 2001 by FAO (FAO 2000a and FAO 2001a), after NI values partly based on the author's own estimates had already been calculated and used for analyses in the present study. The consistency of the author's own estimates of undernourishment for countries in transition for 1995-1997 and the FAO figures for 1996-1998 is evaluated in Chapter 7.2.1.

[6] Whereas the average minimum dietary energy requirements could have been calculated in principle for all countries where the composition of the population by age and sex is known - but only with considerable effort in collecting these data and doing the computations - in many countries no data exist about the distribution of income and/or consumption. Therefore, average dietary energy requirements were omitted in the model, and data about food or income distribution could not be relied on, if the estimation model was to fulfil its purpose of enhancing data availability.

[7] An attempt was made to employ GNP per capita, its square and logarithm in the model. However, respective variables had no additional explanatory power as compared to dietary energy supply per capita and its non-linear transformations.

with the variables referring to the years 1979-81 and 1990-92, respectively. The differences found between the coefficients in the three models that are based on different periods of time, and in the plots of the predicted values against dietary energy supply per capita were marginal (see Figure 4.1.1-1).

Figure 4.1.1-1 Comparison of estimation models based on different periods of time

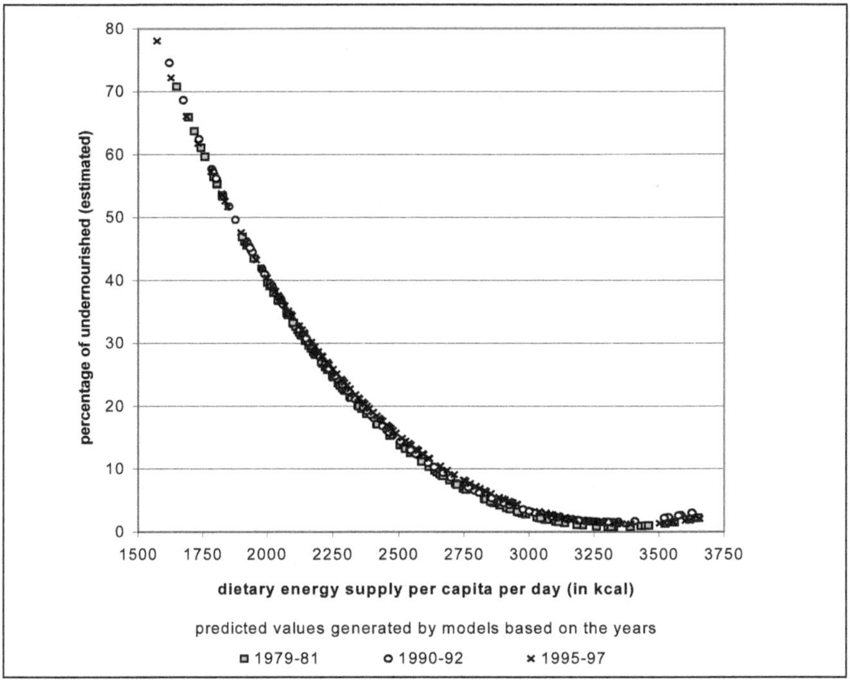

Source Wiesmann et al. 2000, based on data from FAO 1999a and 1999b

To minimise the influence of random errors and to make sure that the final model is derived only from more recent data that are assumed to be of better quality than the 1979-1981 figures, country data for the years 1990-92 and 1995-97 were combined in one data set for the final linear regression[8], again with dietary energy supply per capita

[8] This was done in such a manner that nearly every country in the data set was represented twice, that is by data from the 1990-92 *and* the 1995-97 period. No weighted regression technique was applied, but ordinary least squares, because 2 observations (1990-92 and 1995-97) were available for 98% of the countries in the sample and just 2 countries (2% of the sample) were represented only by one ob-

and its logarithm as independent variables and the percentage of undernourished according to FAO as a dependent variable. The characteristics of this final model are shown in Table 4.1.1-1.

Table 4.1.1-1 Model to estimate the percentage of undernourished, Ordinary Least Squares (OLS)

Dependent variable:	percentage of undernourished		
Independent variables:	dietary energy supply per capita per day (in kcal) (DES), logarithm of DES		
Periods of time:	1990-92 and 1995-97, dependent and independent variables matched		
Variables	**Coefficients**	**T-statistics**	**P-Value**
Intercept	2426.57	30.28	0.0000
DES	0.10	21.07	0.0000
Ln DES	-340.61	-28.88	0.0000
Number of observations	192		
Multiple R	0.986		
R-squared, adjusted	0.971		
Standard error	2.90		
F statistic	3234.55		0.0000
Durbin-Watson statistic	1.96		

Source Wiesmann et al. 2000, based on data from FAO 1999a and 1999b
Note Significance is two-tailed for both F and T-statistics

The significance of the overall model and of the singular coefficients and the constant proved to be very high, which is demonstrated by the significance levels given in Table 4.1.1-1. The probabilities that the whole model is invalid or that any of the coefficients or the constant is zero amounts to less than 0.005% [9]. Dietary energy supply per capita and its logarithm, chosen as independent variables, predict 97% of the variation of the dependent variable (percentage of undernourished), which means that the model has an excellent goodness of fit. Of course, this is not surprising, given the fact that FAO derives the percentage of undernourished mainly from dietary energy supply per capita and assumes a log-linear distribution of dietary energy consumption. The re-

servation from 1995-97. Therefore, it is extremely unlikely that different numbers of data points might introduce a bias into the model.

[9] The Durbin-Watson statistic indicates the absence of auto-correlation, which is not expected to be a problem, because the model is based on cross section data with two observations per country. In this case, the Durbin-Watson statistic shows that the model is not misspecified in terms of variable selection and functional form.

siduals produced by the model can be explained by the impact of specific countries' food distribution data differing from the standard log-linear distribution on the percentage of undernourished and the variation of average minimum dietary energy requirements. Though, as the residuals vary between a minimum of –6.6 and a maximum of 8.2, and the standard error is only 2.9, the influence of the latter factors on the FAO estimates is obviously quite small[10].

The percentage of undernourished turns out to be mainly an inverse, regressive transformation[11] of dietary energy supply per capita, see Figure 4.1.1-2. Discounting is indeed plausible, because declining marginal returns of rising dietary energy supply per capita to the nutrition situation are likely (e.g., in a cross country analysis of child malnutrition, Smith and Haddad 2000 found decreasing marginal returns of DES for children's nutritional status). The upward slope between 3360 and 3600 kcal per capita per day in Figures 4.1.1-1 and 4.1.1-2 is likely to be a bias introduced by the fact that cultural traditions in some Near Eastern states like Turkey foster a high dietary energy supply per capita at a moderate level of development while social inequalities persist. In contrast to that, in a country like South Korea, traditional dietary patterns tend to keep dietary energy supply at about 3,155 kcal per capita per day (in 1997) - a level that is still relatively modest taking into account Korea's economic development - while the percentage of undernourished is very low, partly due to a well performing social system. Consequently, the model designed here should not be used to estimate the percentage of undernourished for developed countries with high dietary energy supply per capita, because the results would be affected by the distortion of the predicted values in the upward slope. Yet, the predicted values for countries with a dietary energy supply over 3,300 kcal per capita per day are irrelevant here anyway. After all, the original data set was supplemented by the author's own estimates for 27 countries that were generated by the model. A complete list of countries with the percentage of undernourished given by FAO and the author's supplementary estimates is given in Table A 5.1-1 in the Appendix.

---

[10] If the predicted values were used for the aggregation of indicators instead of the actual figures published by FAO, the error introduced into the NI would be less than one percentage point for 73% of the countries in the sample (this is the case for all residuals smaller than 3, see also derivation of weights in Subchapter 4.3 - the percentage of undernourished accounts for one third of the NI).

[11] The inversion is evident from the negative slope of the graph (the percentage of undernourished decreases with rising DES). Because of the regressive transformation – i.e., discounting – of DES, the graph flattens out as it goes right (the reduction of the percentage of undernourished for a given increase in DES is the smaller, the higher the level of DES).

Figure 4.1.1-2 Percentage of undernourished, actual and predicted values and residuals plotted against dietary energy supply per capita: model with pooled data (1990-92 and 1995-97)

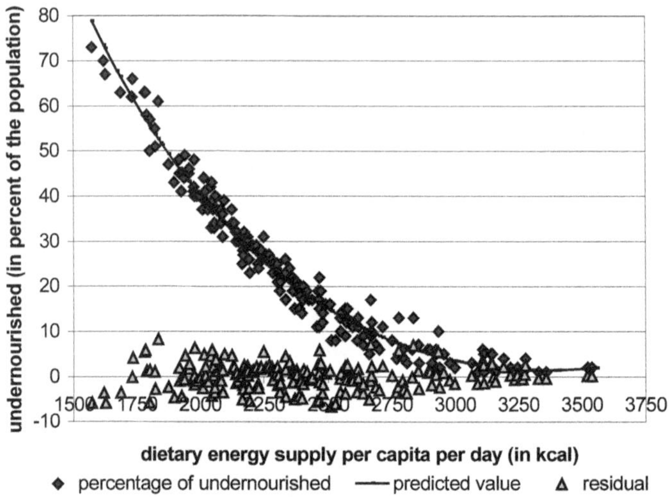

Source Wiesmann et al. 2000, based on data from FAO 1999a and 1999b

4.1.2 Prevalence of Underweight in Children

It is now widely accepted that up to seven years, children all over the world have much the same growth potential. The percentage of underweight children is therefore defined as the proportion of children whose weight-for-age is more than two standard deviations of the reference population below the median value of the reference population (WHO 1997). For the anthropometric data, limited availability is a major problem. They were taken from the WHO's regularly updated database (WHO 1997 and 2000a) and, for the period before 1993, partly from data by the International Food Policy Research Institute (IFPRI) published in the Second Report on the World Nutrition Situation (UN ACC/SCN 1993). The main problem is that the nationally representative surveys these data are drawn from are conducted irregularly, and that for some countries no data are available at all. The data set for underweight children was therefore supplemented by estimates published by UN ACC/SCN in 1993. Care was taken not to include estimated figures that were obviously at odds with existing data, e.g. predicted values with residuals of 10 and more. No estimates of the proportion of underweight children are available so far for the more recent years 1993 to 1998, but fortunately, a great number of surveys providing up-to-date information have been carried out in this period.

Besides the limited availability of data for given periods of time, another problem is posed by the fact that the anthropometric data reported in the nutrition surveys refer to different age groups of children (though the most frequently surveyed group are children under five years). The prevalence of underweight varies between groups of children at different age. For example, nutritional status is usually quite good up to the age of six months, but dramatic deteriorations can be observed after the weaning phase due to inadequate dietary composition and feeding practices. For this reason, the comparison of the prevalence of underweight in children under five years in one country with the prevalence found for the age group 0.25-<3 years in another country, for example, may overestimate the relative severity of malnutrition in the latter. Similarly, the intertemporal comparability of survey results based on different age groups can be affected (the left side of Table 4.1.2-1 shows the age groups used in the surveys that served as data sources for the NI and their share in the overall number of surveys).

This problem can be taken into account by multiplying the data referring to age groups other than the under-fives with corresponding correction factors, as done in the Second World Report on the Nutrition Situation (UN ACC/SCN 1993)[12]. For the purpose of calculating the Nutrition Index, the corrected data as given in the UN ACC/SCN report for the period before 1993 were adopted in this study, ensuring consistency with the estimates taken from the same publication. To improve data consistency over time and internationally, new factors were calculated from existing surveys for which malnutrition prevalence was reported separately for different age groups (data were taken from WHO 2000a). In the selection of nationally representative nutrition surveys for this purpose, preference was given to countries with nutrition problems where underweight prevalence has also been reported for age groups other than the under-five-year olds. Special emphasis in this group was put on surveys from countries where proportions of underweight children higher than 6% are known. The latter selection criterion takes into account that errors introduced into the NI by data referring to the 'wrong' age group might be the larger in absolute terms the higher the underweight prevalence[13], with larger distortions of the NI scores as a result.

---

[12] As these factors are found to be different for each country and level of malnutrition, it is not clear how their numerical values have been determined.

[13] In the Second Report on the World Nutrition Situation, the factors are reported to be in the range of 1.01 to 1.30. For an underweight prevalence of 6%, errors below 0.9 % points can be expected.

Table 4.1.2-1 Frequency distribution of reference age groups before and after conversion of malnutrition prevalence (in age groups different from 0 - < 5 years) into the prevalence in under-five-year olds

Reference Age Group	Before Correction *		After Correction **	
	Number of Surveys	Percentage of Surveys	Number of Surveys	Percentage of Surveys
0 - < 3 years	18	10.4	12	6.9
0.25 - < 3 years	6	3.5	1	0.6
0 - < 4 years	4	2.3	0	0.0
0.25 - < 4 years	1	0.6	0	0.0
0.25 - < 5 years	4	2.3	2	1.2
0.5 - < 5 years	11	6.4	0	0.0
0 - < 6 years	5	2.9	0	0.0
0.5 - < 6 years	2	1.2	0	0.0
1 - < 6 years	1	0.6	0	0.0
1 - < 7 years	1	0.6	1	0.6
all other than 0 - < 5 years	**53**	**30.6**	**16**	**9.2**
0 - < 5 years	**120**	**69.4**	**157**	**90.8**
All age groups	**173**	**100.0**	**173**	**100.0**

Source author's own calculation, based on data from WHO 2000a and UN ACC/SCN 1993

Notes * As reported in the nutrition surveys (WHO 2000a; UN ACC/SCN 1993).
** Surveys where no correction factor was applied, because prevalence in the respective age group proved to be almost identical with prevalence in under-five-year olds, were attributed to the 0 - < 5 years group.

The correction factors for a specific country and survey were obtained by dividing the underweight prevalences for the age groups in question (0 to under 3 years, 0.5 to under 3 years, 0 to under 4 years, and 0.5 to under 5 years) by the prevalence reported for all under-fives[14]. The box-plots in Figure 4.1.2-1 show the range, median and quartiles of correction factors found for different age groups. Evidently, the variation of factors among countries is high for the two age groups of under-three-year olds (0-<3 years and 0.5-<3 years) and very low for the under-four-year olds and the 0.5-<5 group, see Figure 4.1.2-1.

[14] The prevalence of malnutrition in under-three-year olds, for example, is usually not directly indicated in the data source, but could be calculated by aggregating the prevalences in the 0 - <0.5, 0.5 - <1, 1 - <2 and 2 - <3 age groups (weighting the percentage of underweight children with the sample size in each age group).

Figure 4.1.2-1 Box-plots of correction factors (prevalence of malnutrition in indicated age group divided by prevalence in all under-five-year olds)

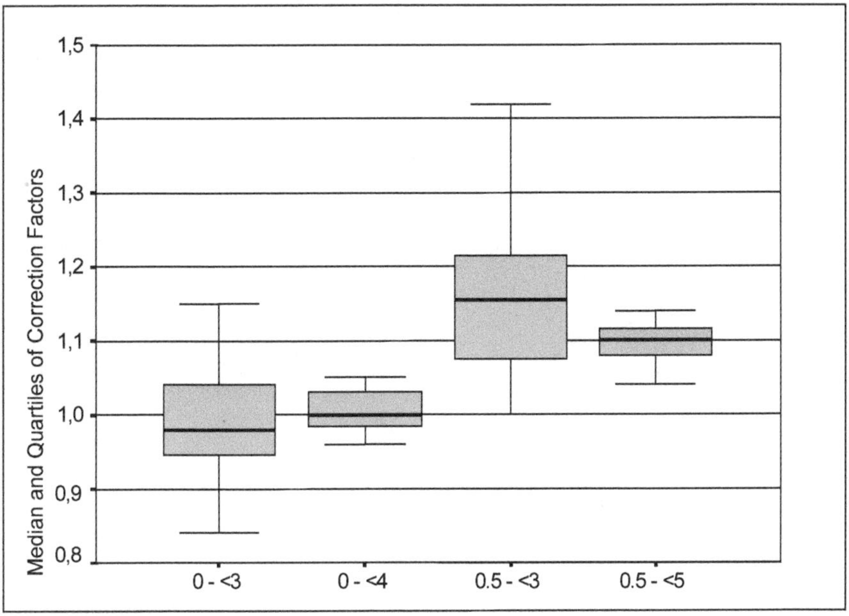

Source Wiesmann et al. 2000, based on data from WHO 2000a

Notes The box covers the middle 50% of the data values between the upper and lower quartiles; the central line is the median. 'Whiskers' extend to the highest and lowest values, unless these are outliers (at a distance of more than 1.5 times the interquartile range from the margin of the box) or extreme values (at a distance of more than 3 times the interquartile range from the margin of the box).

Correction factor patterns prove to be astonishingly stable over time for a given country, at least for countries where nutrition surveys are available for several years. A small time series of factors could be calculated for these countries (compare Figure 4.1.2-2 and Table A 4.1.2-1 in the Appendix, which also reveals regional peculiarities of the South and South-East Asian region).

Figure 4.1.2-2 Change of correction factors over time, country cases

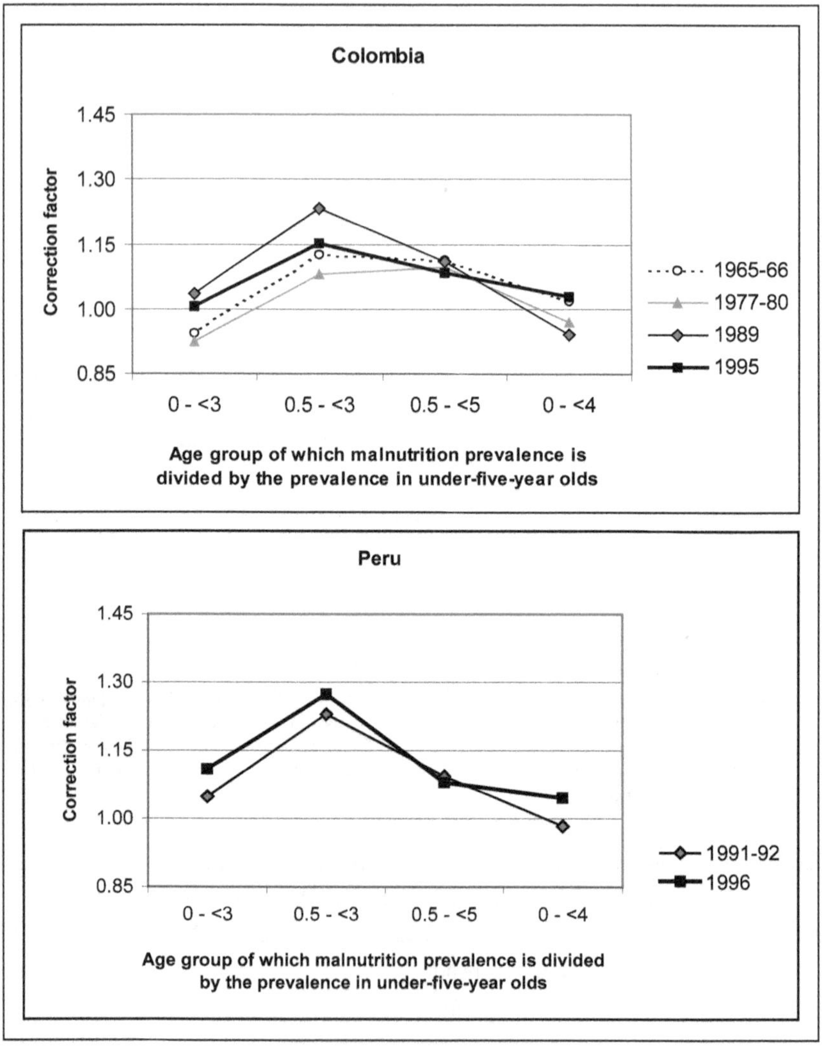

Source Wiesmann et al. 2000, based on data from WHO 2000a

Figure 4.1.2-2 (continued) Change of correction factors over time, country cases

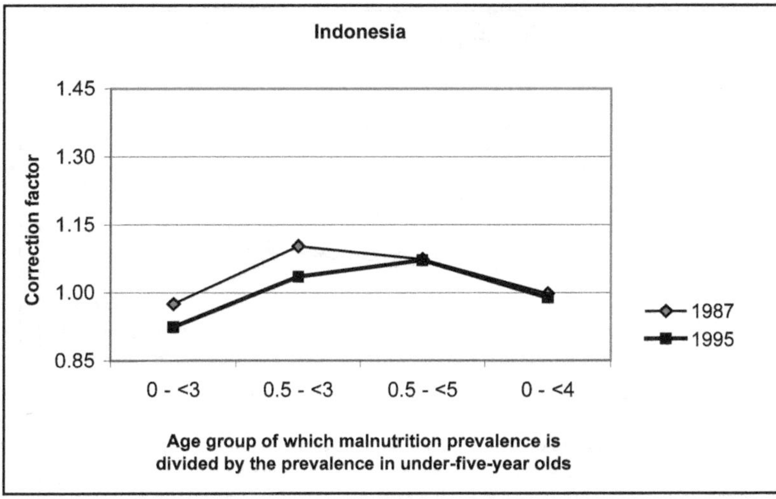

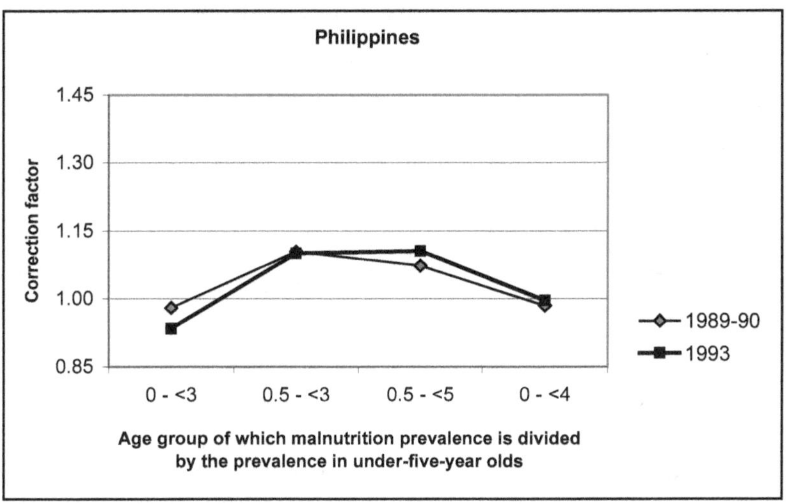

Source Wiesmann et al. 2000, based on data from WHO 2000a

Figure 4.1.2-2 (continued) Change of correction factors over time, country cases

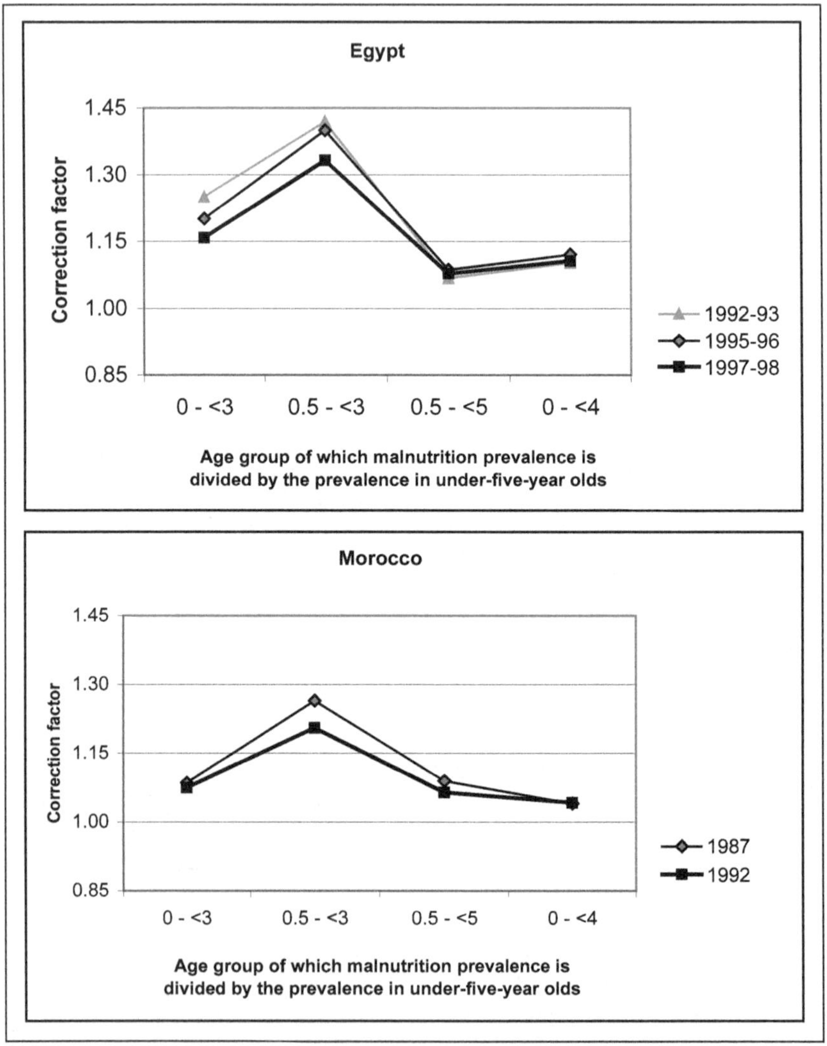

Source Wiesmann et al. 2000, based on data from WHO 2000a

Figure 4.1.2-2 (continued) Change of correction factors over time, country cases

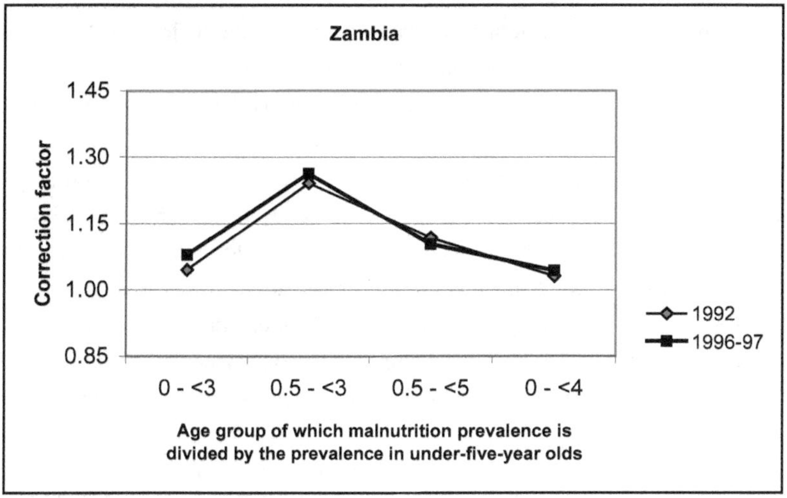

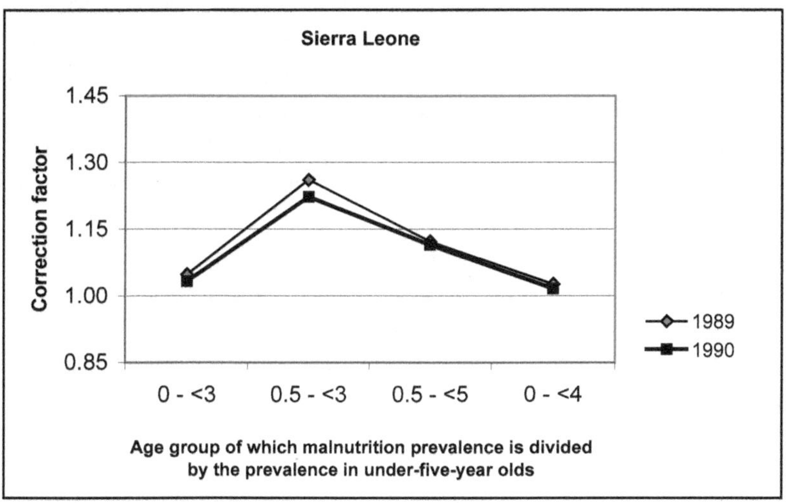

Source Wiesmann et al. 2000, based on data from WHO 2000a

In conclusion, the procedure described below was followed to improve data consistency:

- No correction factors were applied to data referring to the under-four age group, because the mean value of correction factors turned out to be one and the standard deviation proved to be very low, making it unlikely that large deviations from the underweight prevalence in under-fives might occur.
- Correction factors were always applied to data referring to the 0.5–<5 group, because the calculated factors showed to be consistently greater than one with low standard deviation. In cases where a correction factor could be derived for the same country at a different point in time, this factor was used. If this was not possible due to lack of data, the average value of factors from countries of the same region was taken. Where neither factors for the respective country nor for its region could be calculated, the average of all factors was chosen (only in one case, for Azerbaijan).
- Taking into account the high variation of correction factors for the 0–<3 and 0.5-<3 age group even within regions, and their average value of one, only factors derived from data for the same country were applied to avoid the enlargement of an existing bias with the real, yet unknown prevalence among under-fives[15]. Therefore, correction was impossible for some countries, which does not necessarily mean that the underweight prevalence finally used for NI calculation deviates from the real prevalence among under-five-year olds (because the unknown correction factor could also be exactly or nearly one).
- For some countries, anthropometric data referred to the 0.5–<6 age group. In this case, first the prevalences for the 0.5–<5 age group were calculated from the age-specific prevalences reported in the survey. Then, the correction factor for this group was applied to estimate the probable under-five prevalence. Differences found between prevalences in the under-five and under-six age groups were so small that adjustment was considered unnecessary.

After this data refinement, more than 90% of the survey data expressed the underweight prevalence in the under-five age group (see Table 4.1.2-1, right hand side).

[15] Correction was omitted if the correction factor for a specific country was in the range of 1+/- 5%.

4.1.3 Under-Five Mortality Rate

The under-five mortality rate indicates the mortality risk between birth and age five, and is usually given in deaths of children under five years per 1000 live births (UNICEF 1999). The under-five mortality rate is available for almost every country in the world and for a broad range of years, since it is published annually by UNICEF in the reports on the State of the World's Children (SOWC). The data are derived from official statistics or collected in surveys (such as the Demographic and Health Surveys), and partly estimated for the SOWC reports. They are taken from UNICEF (1995) and UNICEF (1999) for the purpose of NI calculation. Although this is an unusual way to express mortality rates, the under-five mortality rate is transformed to the percentage of deaths up to the age of five for a given total of children born for the purpose of index calculation. This linear transformation is achieved by simply dividing the under-five mortality rate (per 1000 live births) by ten, and is carried out in order to adapt this indicator to the metric of the other two NI components.

Underlying data for the calculation of subnationally disaggregated NI values for China and India, including the under-five mortality rate by province or federal state, are discussed in Boxes 4.1.3-1 and 4.1.3-2.

Box 4.1.3-1 Underlying data of NI components for China's Provinces

To calculate NI values for China's provinces, in addition to information on child malnutrition and mortality disaggregated by province, data on the extent of undernourishment are needed. No such estimates have been published by FAO on a subnationally disaggregated level. Again, the estimation model for the percentage of undernourished described in Subchapter 4.1.1 can be applied to fill the data gaps. Food consumption statistics published in National Bureau of Statistics of China 1999, for rural households are used to assess dietary energy intake per capita, which is a necessary input for the model. The food categories given in this publication are very broad, e.g., 'grain' or 'pork, beef and mutton', and the exact proportions of the various foodstuffs assembled in one category are not made explicit. Data from the FAO's Food Balance Sheet for China in 1998 (FAO 2000b) are taken to derive the average dietary energy content per category in kcal per kg[16].

[16] This is done by inserting the figures indicated in China's Food Balance Sheet for 1998 (FAO 2000b) for a category into the following simple formula:

ADEC(X) =	DESC(X)*365/FS(X) when
ADEC(X) =	Average Dietary Energy Content of Category X (in kcal per kg)
DESC (X)	Dietary Energy Supply from Category X per capita per day (in kcal)
FS(X)	Food Supply of Category X per capita per year (in kg)

For sugar and edible oil, dietary energy content figures have been directly adopted from nutritive value tables (Elmadfa et al. 1988), see Table A 4.1.3-1-1 in the Appendix for details.

Box 4.1.3-1 (continued) Underlying data of NI components for China's Provinces

The obtained figures, which are shown in Table A 4.1.3-1-1 in the Appendix, are representative for the average Chinese diet. However, variations across provinces in the composition of categories of foodstuffs cannot be considered by this method. The dietary energy intake figures shown in Table B 4.1.3-1-1 for the rural population by province should therefore be regarded as approximations. They are the basis for estimating undernourishment by province for rural areas.

Yet, the same method of estimating the percentage of undernourished as for rural areas is not applicable to urban areas, or to the total population. The reasons can be found in the specific dietary energy consumption patterns observed for China's urban population. First, according to food consumption statistics, urban households tend to consume less dietary energy than rural households, despite their much higher average incomes (compare income data for the rural and urban population in Table B 4.1.3-1-1). Except for Hebei, the dietary energy consumption per capita of all income groups is higher for the rural than for the urban population (Chunming and Zongming 1994). This tendency is also confirmed by the results of preliminary calculations with more recent food consumption data of urban households (1998). Explanations that can be thought of are higher dietary energy needs of rural people involved in heavy physical work, or failure to properly capture the increasing consumption of food outside the home in urban areas in consumption and expenditure surveys. Second, whereas dietary energy intakes of the rural population tend to rise steeply from the lowest to the highest income group, stabilisation of dietary energy consumption at a relatively low level, or even a relative drop in dietary energy consumption of the wealthiest group can be observed for urban people (Chunming and Zongming 1994).

Both phenomena, if not caused by biases in data collection, are in contradiction to the general observation of dietary energy intakes rising with increasing wealth. If people reduce their dietary energy intakes as they become better off - which is surely not to be interpreted as increasing inability to meet their dietary energy needs (rather as a reduction of such needs) - the model based on dietary energy supply per capita is no longer applicable. Due to their construction, the FAO method and the model approximating its results would inevitably misinterpret a reduction in dietary energy consumption of the wealthiest and a consequent shift of the total average to the left as increasing undernourishment. In contrast to the situation in India, where the majority of inhabitants in each Federal State still live in rural areas, in several Chinese provinces the proportion of the urban population is more than 50%. Hence, the relatively low dietary energy intake in urban areas as calculated from food consumption data (be it due to measurement biases or for other reasons) produces irritating results even for a province's total population if the estimation model is applied imprudently. One example is the assessment of the highest level of undernourishment in China for Beijing, which has by far the highest share of urban residents (Table B 4.3.1-1-1).

Box 4.1.3-1 (continued) Underlying data of NI components for China's Provinces

Table B 4.1.3-1-1 Dietary energy intake, population and income in Chinese Provinces

	total population[1] (in mio) 1998	urban population[2] (in %) 1997	dietary energy intake per capita, rural[3] (in kcal) 1998	net income, rural[1] (in Yuan) 1998	net income, urban[1] (in Yuan) 1998
Anhui	62.4	25.3	2501	1863	4770
Beijing	12.6	82.7	2070	3952	8472
Chongqing	30.9	39.7	2630	1720	5467
Fujian	33.2	27.5	2763	2946	6486
Gansu	25.5	24.5	2450	1393	4010
Guangdong	71.7	34.1	2777	3527	8840
Guangxi	47.2	18.1	2467	1972	5412
Guizhou	36.7	23.3	2387	1334	4565
Hainan	7.6	26.6	2412	2018	4853
Hebei	66.3	21.7	2114	2405	5085
Heilongjiang	38.1	54.7	2752	2253	4269
Henan	93.9	19.2	2345	1864	4219
Hubei	59.5	31.2	2988	2172	4826
Hunan	65.7	25.9	3185	2065	5434
Inner Mongolia	23.7	35.9	2716	1981	4353
Jiangsu	72.5	29.2	2770	3377	6018
Jiangxi	42.3	27.3	3034	2048	4251
Jilin	26.7	45.7	2543	2384	4207
Liaoning	42.0	53.1	2626	2580	4617
Ningxia	5.4	28.3	2620	1721	4112
Qinghai	5.1	26.9	2619	1425	4240
Shaanxi	35.9	26.0	2248	1406	4220
Shandong	89.1	27.6	2401	2453	5380
Shanghai	14.7	78.4	2641	5407	8773
Shanxi	32.0	34.8	2164	1859	4099
Sichuan	85.8	35.1	2616	1789	5127
Tianjin	9.7	70.4	2310	3396	7111
Tibet	2.5	10.9	2366	1232	..
Xinjiang	17.5	39.8	2551	1600	5001
Yunnan	42.0	17.4	2569	1387	6043
Zhejiang	45.0	35.5	2676	3815	7837
China total*	1243.4	31.3	2585 **	2162	5425

Source author's own presentation, based on data from National Bureau of Statistics of China 1998 and 1999, FAO 2000b

Notes 1 National Bureau of Statistics of China 1999
2 calculated from population data published in National Bureau of Statistics of China 1998
3 calculated from food consumption data of rural households published in National Bureau of Statistics of China 1999
* The listed 31 provinces comprise 100% of the Chinese population (1998)
** 2973 kcal per capita per day according to FAO's Food Balance Sheets (1998)

Box 4.1.3-1 (contin.) Underlying data of NI components for China's Provinces

Undernourishment in China's urban population thus has to be estimated differently. The percentage of people existing below the poverty line, which is usually defined by the minimum income needed to afford a bundle of basic necessities (including food), may serve as a yardstick. Unfortunately, data on the incidence of poverty are not available by province for the urban population. Aggregated figures for the total urban population are listed in Sheng 2002 for the years 1991 to 1996. If the trend of a decline stated for 1996 has persisted to 1998, approximately 3% of urban households were absolutely poor in 1998, lacking food or other basic necessities. Accordingly, a uniform proportion of undernourished, 2% in the provinces' urban areas, is presumed in the following - which is of course a crude assumption. This figure for urban households is aggregated with the percentage of undernourished in the rural population estimated from the model by weighting with the respective population shares[17].

For the malnutrition and child mortality data, reliability is not guaranteed either. The respective data have been compiled by provincial statistical bureaux and reported to the government for the years 1990, 1995 and 1998. The data were not gained from a survey. However, comparison with anthropometric survey data of eleven provinces collected in 1987, 1990 and 1992 (WHO 2000a) shows approximately consistent patterns and trends in underweight in children. Aggregation of official provincial data for underweight in children gives a prevalence of 7.3%, which is less than the 9.0% calculated from the anthropometric survey in 1998, but quite close to the survey result[18]. Similarly, the under-five mortality figure aggregated from the provincial data is with 38 per thousand live births slightly lower than the number published by the Ministry of Health drawn from the Child Mortality Surveillance System, which reports 42 deaths per thousand live births (for the purpose of aggregation, the number of live births by province has been calculated from the birth rates and total provincial population with data from National Bureau of Statistics of China 1999). UNICEF has published an even higher estimate for 1998, (47 per 1000), which is extrapolated from survey data in the early 1990s (UNICEF 2001). After all, whereas the provincial child mortality and malnutrition figures shown in Table B 5.1-1-1 (see Chapter 5.1) are surely not of perfect quality, the above cited checks indicate that they are at least not totally misleading.

---

[17] This approach has obvious weaknesses, but in the absence of more information, it is a 'second best solution'. For one province – Hebei – the estimate of undernourishment in the rural population, that amounts to 32.3%, is considered much too high. The model devised in Section 4.1.1 cannot take into account differences in variation of dietary energy intake, and is therefore probably not suitable for rural Hebei, which seems to have a narrower distribution of dietary energy intakes than other rural areas (Chunming and Zongming 1994).

[18] Since there are large differences in the age composition of the population across Chinese provinces - the percentage of children up to the age of 6 years varies from 4% in Shanghai to 12% in Tibet - the number of children 6 years and younger has been calculated by province from data in the China Population Statistics Yearbook 1998 (the proportion of children under five was not available), and these numbers have been used for the weighted aggregation of child malnutrition data.

Box 4.1.3-2 Underlying data of NI components for India's Federal States

The NI values for Indian states are obtained or estimated from various sources of state level data. In order to derive the percentage of undernourished from data on dietary energy intake per capita by federal state, the estimation model presented in Section 4.1.1 is applied. Data on dietary energy intake are used instead of data on dietary energy supply as an input for the model. They have been calculated from food consumption data of the 50th Round of the National Sample Survey (1993-94) by Indian statisticians for urban and rural areas separately. The data for the rural and urban population are aggregated for each state by means of the percentage of the urban population, which is calculated from census data referring to 1991. The population data from the 1991 Census are also used to calculate aggregate dietary energy intake per capita from the state figures for India as a whole (96% of India's total population in 1991 resided in the 15 major states for which dietary energy consumption data are available), see Table B 4.1.3-2-1.

Table B 4.1.3-2-1 Dietary energy intake and population in Indian States

	total population[1]	urban population[1]	dietary energy intake per capita, unrefined			dietary energy intake per capita, upgraded
			rural[2]	urban[2]	total[3]	total[3]
	(in mio)	(in %)	(in kcal)	(in kcal)	(in kcal)	(in kcal)
	1991	1991	1993-94	1993-94	1993-94	1993-94
Andhra Pradesh	66.5	26.9	2052	1992	2036	2131
Assam	22.4	11.1	1983	2108	1997	2090
Bihar	86.4	13.1	2115	2188	2125	2224
Gujarat	41.3	34.5	1994	2027	2005	2099
Haryana	16.5	24.6	2491	2140	2405	2517
Karnataka	45.0	30.9	2073	2026	2058	2154
Kerala	29.1	26.4	1965	1966	1965	2057
Madhya Pradesh	66.2	23.2	2164	2082	2145	2245
Maharashtra	78.9	38.7	1939	1989	1958	2050
Orissa	31.7	13.4	2199	2261	2207	2310
Punjab	20.3	29.5	2418	2089	2321	2429
Rajasthan	44.0	22.9	2470	1922	2345	2454
Tamil Nadu	55.9	34.2	1884	1922	1897	1985
Uttar Pradesh	139.1	19.8	2307	2114	2269	2374
West Bengal	68.1	27.5	2211	2131	2189	2291
India total*	846.30	25.71	2153	2071	2132 **	2230 **

Source author's own presentation, based on data from Census Data Online 2001 and on personal communication from Mahendra Dev

Notes 1 calculated from data published in Census Data Online 2001

2 personal communication from Mahendra Dev, the figures have been calculated from food consumption data from the 50th round of the National Sample Survey (NSS)

3 author's own calculation

* The listed 15 States comprise 96% of the Indian population (1991)

** 2330 kcal per capita per day according to FAO's Food Balance Sheets (average 1993-94)

Box 4.1.3-2 (continued) Underlying data of NI components for India's Federal States

The initial figures of dietary energy intake obtained for India and its states may slightly underestimate actual dietary energy consumption. The consumption of street food in urban areas, for example, is hard to record in consumption and expenditure surveys. This may partly explain why the figure for dietary energy supply in India as a whole is, according to the FAO, about 10% higher than the figure produced by aggregating state level dietary energy consumption data. On the other hand, the FAO figures for dietary energy supply are likely to overestimate dietary energy intake, because wastage and use of food for animal feed in households are difficult to assess at the national level. The state data for dietary energy intake are, therefore, increased by 4.66%, allowing the aggregate figure for India to reach the average formed by the aggregated, uncorrected state data (2132 kcal per capita per day) and the FAO figure for 1993-94 (2330 kcal per capita per day, FAO 2001a). The upgraded dietary energy intake figures, that are inserted into the estimation model for the percentage of undernourished, are shown in Table B 4.1.3-2-1 together with the original dietary energy intake and population data. Again, aggregation of state figures for undernourishment is carried out for a cross-check with FAO figures. FAO estimates undernourishment in India at 26% in 1990-92 and 22% in 1995-97 (FAO 1999a). Aggregating the estimates of undernourishment of the 15 states for 1993-94 produces a number of 25.8 % for India as a whole, which fits the trend indicated by the FAO estimates very well. This is the case despite the relatively higher dietary energy supply figures on which the FAO estimates are based (2335 kcal per capita per day in 1990-92). The reason for this can be found in the fact that the model used here tends to underestimate the percentage of undernourished in India by 5 to 6 percentage points. This is actually compensated for by relying dietary energy intake figures that are slightly lower than the dietary energy supply figures published by FAO.

Child malnutrition and under-five mortality data are drawn from the National Family Health Survey in India in 1992-93. Astonishingly, the correlations between the percentage of undernourished and both child malnutrition and mortality are weak and insignificant, and for the latter even negative (rank correlation coefficients of 0.01 and –0.28, respectively). In contrast, the rank correlation between the two indicators referring to children is about the same as for international data (0.54, p-value = 0.038). This is not necessarily an argument against the validity of the dietary energy intake data, because cultural and social influence factors may gain importance within the limited range of subnationally disaggregated child malnutrition and mortality data. In a study in Ethiopia, Pelletier et al. found that child nutritional status was only weakly associated with proxies of household food availability (Pelletier et al. 1995). Though data on child malnutrition refer to the age group of under-four-year olds and not to under-five-year olds, no adjustment is made in accordance with the criteria for applying correction factors that have been laid out in Subchapter 4.1.2 (the underweight prevalence in under-four-year olds differs very little from the prevalence in under-fives, which also holds for India as a whole, see Table A 4.1.2-1 in the Appendix).

4.2 Reference Periods for NI Calculation

Gaps in available data could be filled and international comparability improved by estimation and refinement procedures. Nevertheless, the data on undernourishment and the anthropometric data cannot be matched perfectly because they do not refer to the same years. In particular, the anthropometric data are spread over a wide time range. Therefore, anthropometric data for NI calculation cannot be drawn from one year, but the latest available information out of a fixed period of time has to be used. Justification can be found in the fact that rapid changes in the percentage of underweight children rarely occur. There is a built-in lag in this indicator, partly due to the fact that the eldest children in the sample are usually nearly five years old, and their weight for age is not only determined by current conditions, but also the outcome of their nutrition situation since birth and even before birth[19]. Moreover, some factors that have been identified as relevant for children's nutritional status – for example female education and access to safe water (Smith and Haddad 2000) – surely require long-term investments before showing an impact at a nationwide level.

There is also empirical evidence for rather slow changes in underweight prevalence (compare the list of countries with more than one nutrition survey in the 1965-97 period in WHO 1997, pp. 40-42, where information on the year of the survey, reported prevalence and annual change rates for the time between the surveys is given). About 65% of the reported annual change rates are in the range of –1 to +1% points, and nearly 80% in the range of –1.5 to +1.5% points. Only 7% of all observed change rates – a total of six cases - amount to an absolute value of more than 4 percentage points per year. In this context, it should be noted that in five out of the six cases with relatively high annual change rates, these rates have been calculated from data out of surveys that were completed in subsequent years. Because inevitable measurement and sampling errors gain the greater influence on calculated annual change rates the shorter the time periods between the surveys, these five rates are likely to be more distorted by random errors than the change rates computed from surveys with larger time intervals[20].

---

[19] This holds even more for the height-for-age indicator (stunting), because there may be some weight recovery for underweight children up to five years, whereas impaired growth is difficult to reverse (UN ACC/SCN 1997). However, underweight in children is the preferred anthropometric indicator, because - among other reasons - the proportion of stunted children may be too time-resistant for indicating changes without considerable delay.

[20] The following fact supports this assumption: in all three out of the five countries for which more than one annual change rate could be calculated, the trends computed before or after the one-year pe-

Taking into consideration that annual change rates in underweight prevalence are quite small for the majority of countries, six-year periods have been chosen as the reference periods from which the latest available information about underweight prevalence in children was drawn for NI calculation. Furthermore, the bulk of data, including the estimates, refer to the more recent years in these periods (compare Figure A 4.2-1 in the Appendix), e.g. more than three quarters of the data collected in the 1993-1998 period are from the years 1995-1998. Therefore, the latest information from 1993-1998 is considered approximately representative for the year 1997, for example. Similarly, the percentage of undernourished referring to 1995-1997 is indicative for the year 1997. The under-five mortality rate has been directly taken from 1997. The complete listing of time ranges or years which the single indicators refer to, and the respective NIs, are found in Table 4.2-1.

Table 4.2-1 Time ranges for NI calculation

Reference Year of Nutrition Index	Corresponding Reference Years and Periods of the NI Components
NI 1981	Undernourished 1979-1981 Underweight children, data from most recent survey in 1977-1982 [1] Under-five mortality rate 1980
NI 1992	Undernourished 1990-92 Underweight children, data from most recent survey in 1987-1992 [1] Under-five mortality rate 1992 [2]
NI 1997	Undernourished 1995-97 Underweight children, data from most recent survey in 1993-1998 [1] Under-five mortality rate 1997

Source author's own presentation

Notes 1 The year of survey completion is defined as the 'reference year' here.
2 For some small countries – Comoros, Djibouti, the Gambia, Guyana, and Swaziland – data for 1992 were not available and data from 1993 were taken instead.

Apparently, the time intervals between the reference periods for data about undernourishment entered into the NI 1981, 1992 and 1997 are equal for all countries, and the same applies to the under-five mortality rate[21]. In contrast to this, the intervals between the survey data on malnutrition included in the NI 1981, 1992 and 1997 differ

riod with a change rate above 4% points go in the opposite direction (as indicated by the opposite sign of the change rate).

[21] With the negligible exception of the five countries mentioned in Table 4.2-1, for which the under-five mortality data of 1993 were used.

between countries. This can be seen as a problem for the interpretation of intertemporal changes in NI scores. The interval between the underweight data used for the NI 1992 and 1997 may in some extreme cases be only one year for some countries and up to 10 years for others. The expected interval is about six years, and 10 years between the data of the 1977-82 and 1987-92 reference periods. The time intervals between the nutrition survey data and estimates were analysed: the average of the intervals derived from the data in the reference periods for the NI 1992 and 1997 amounted to 5.2 years, and the mean of the corresponding intervals referring to the NI 1981 and 1992 was 9.9 years. The standard deviations both amounted to 1.8 years. Indeed, maximum and minimum values of 10 years and one year, respectively, were observed for the intervals between underweight data in the 1987-92 and 1993-1998 periods, but they applied only to three out of 65 countries, and 83% of intervals were found in the quite reasonable range of seven to three years. Similarly, the extreme values of intervals calculated for the data from 1977-82 and 1987-92 were found to be five and 13.5 years, but again, about 80% of the intervals were in a quite sensible range between eight and twelve years.

In conclusion, it may be stated that the problem of differing time intervals for the underweight data exists and is not yet resolved, but that the magnitude of the problem is not large with respect to the validity of the NI – partly due to the attenuating effect of constant time intervals for the other two indicators. Solutions for the problem can be expected from a higher frequency of nutrition surveys and more reliable prediction models in the future.

4.3 Standardisation and Aggregation of Indicators

Different subsets of the total data set were subjected to factor analysis, with estimates for the percentage of the undernourished and for underweight in children included one time and excluded another time, and comprising data from several periods of time (1977-1982, 1987-92, 1993-98, pooled data for 1987-98, etc.)[22]. Weights were found to be quite stable at 0.32, 0.33 and 0.35 for underweight children, undernourishment and child mortality, respectively. Because data availability and reliability are better in more recent periods, the statistics and derived weights presented here are based on a combined data set for the 1987-1998 period.

---

[22] Since each dimension of the NI is represented by only one variable, the application of a weighted principal components analysis was not necessary (compare Chapter 2.3.3).

Table 4.3-1 Statistics of principal components analysis

Correlation Coefficients			
	Under-five mortality rate	**Undernourished (in%)**	**Underweight children (in%)**
Under-five mortality rate	1	-	-
Undernourished	0.68***	1	-
Underweight children	0.63***	0.55***	1

*** Coefficients are significant at the 0.001 percent level.

Anti-Image Correlation Matrix			
	Under-five mortality rate	**Undernourished (in%)**	**Underweight children (in%)**
Under-five mortality rate	0.66	-	-
Undernourished	-0.51	0.71	-
Underweight children	-0.41	-0.22	0.76

Notes: Measures of sampling adequacy for variables are printed on the diagonal.
The Kaiser-Meyer-Olkin Measure of sampling adequacy is 0.706.

Factor Statistics			
	Eigenvalue	**Percent of**	**Cumulated**
Factor 1	2.24	74.7	74.7
Factor 2	0.46	15.2	89.9
Factor 3	0.30	10.1	100.0

Note: Only Factor 1 was finally extracted by means of principal components analysis (criterion: eigenvalue > 1).

Final Statistics of Variables			
	Communality	**Factor Score**	**Derived Weight***
Under-five mortality rate	0.80	0.90	0.35
Undernourished	0.74	0.86	0.33
Underweight children	0.70	0.84	0.32

* Formula for Derivation:
Weight of Variable X = Factor Score of Variable X / (Sum of all Factor Scores)

Source author's own calculation

Statistics drawn from factor analysis are shown in Table 4.3-1, including the absolute correlations of the variables they are based on, and their one-tailed significance (estimates of undernourishment and underweight in children are considered in this analysis). The correlations between the variables are highly significant and in the range of medium to high[23]. The Kaiser-Meyer-Olkin measure of sampling adequacy is 0.71 for

[23] These results are in contradiction to the findings of M. Nubé, who questions the validity of FAO's undernourishment estimates on the grounds of low and insignificant correlations with anthropometric indicators for children and adult women (Nubé 2001). Yet, he uses a very limited data set of 23 countries (due to limited data availability on adult nutritional status), 13 of which are Sub-Saharan African countries with probably poorer data quality than countries in other regions. In contrast to his evidence, correlations are still highly significant for the much larger data set employed in the present analysis, even if estimates of underweight in children and undernourishment are omitted. The fact that there is

the selected variables and therefore falls into the category that is usually classified as 'middling' or 'quite good'. Measures of sampling adequacy for single variables that are given on the diagonal of the anti-image correlation matrix indicate that all three indicators are well-suited for a common factor analysis. By means of principal components analysis, one factor was extracted that accounted for 74.7 % of variation. Factor scores obtained for the indicators were divided by their sum in order to normalise the sum of weights to one.

The basic formula for the calculation of the NI is as follows:

$$\text{NI} = 100 - (0.33\ \text{PUN} + 0.32\ \text{CUW} + 0.35\ \text{CM}) \tag{4.1}$$

when NI: Nutrition Index
PUN: percentage of the population undernourished
CUW: percentage of children underweight
CM: percentage of children dying before age five (child mortality)

The better the nutrition situation, the more the NI approximates to its theoretical maximal score of 100. The theoretical minimum value of zero is not reached for countries if the NI is calculated from unstandardised components according to Formula (4.1), because this would indicate that all children died before their fifth birthday, that the whole population was undernourished and that 100% of children were underweight[24]. For the basic version of the NI, no standardisation of index components is carried out prior to aggregation, since the underlying variables of the NI are expressed in percentages, thus sharing a common metric with natural limits. The actual maximum and minimum values for the percentage of undernourished, the prevalence of underweight, the under-five mortality rate and the basic NI are depicted in Table 4.3-2 for 106 countries (NI 1997 supplemented by NI 1992 in case more recent data are not available).

---

even a closer correlation for undernourishment and child mortality than for the anthropometric indicator and under-five mortality may further confirm the relevance of food inadequacy estimated by FAO (see also Chapter 6.1.1 for the results of the rank correlation analysis with cross section data for 106 countries, demonstrating the significant association of NI components, which is not, however, found for some subsamples of the full data set).

[24] This scenario is extremely unlikely for an entire population, though it may be not too far from the sad reality for vulnerable groups like refugees during times of extreme crisis.

Table 4.3-2 Descriptive statistics for the NI and its partial indicators

	Minimum	Maximum	Mean	Standard Deviation	Variance
Undernourished (in%)	1.0	70.0	22.7	17.8	316.0
Underweight in Children (in%)	0.6	57.3	20.6	13.8	190.1
Under-five-mortality (in%)	0.9	32.0	9.6	7.4	54.2
Nutrition Index (NI)	54.8	98.0	82.6	11.4	129.2
Number of cases: 106					

Source author's own calculation, based on data described in Chapter 4.1

Whereas the ranges, maximum values and standard deviations of the percentage of undernourished and underweight in children are roughly similar, they are only about half of the values of these indicators for the under-five mortality rate (expressed as a percentage, not per 1000, see Chapter 4.1.3). The weighting factors derived from principal components analysis, however, are based on correlation coefficients referring to normalised variables. Thus, applying these weights to an unstandardised set of index components may introduce unintentional weighting into the NI (compare Chapter 2.3.3). For this reason, three alternative NI versions have been calculated with partial indicators standardised according to the formulas in Chapter 2.3.2.2. They are tested for redundancy vis-à-vis the basic NI by means of rank correlations. For the first of these NI versions, the actual maximum values listed in Table 4.3-2 for the NI components are inserted into Formula (2.1) for standardisation. For the second version, a minimum value of zero is assumed for the application of Formula (2.2) to the partial indicators, and estimated maximum values of 80% for the percentage of undernourished, 72% for the prevalence of underweight in children and 40% for under-five mortality are fixed as upper limits[25]. For the third, NI components are normalised according to Formula (2.3). The partial indicators standardised in these three different ways are aggregated by means of Formula (4.1). The resulting indices (termed NI stand. (2.1), NI stand. (2.2) and NI stand. (2.3)) are compared to the basic NI calcu-

[25] 80% was chosen for the percentage of undernourished since the share of undernourished had risen to around 70% in some countries by the end of the 1990s (FAO 1999a). An upper limit of 72% is set for underweight in children, because the highest national prevalence since the beginning of systematic data collection was observed in India with 71.3% in 1974-1979 (WHO 1997). 40% is the maximum fixed for under-five mortality, corresponding to the extremely high child mortality rate of 400 per 1000 live births 1960 in Mali (UNICEF 1995).

lated from unstandardised components by means of rank correlation analysis for the total sample and subsamples grouped by NI level, see Table 4.3-3.

Table 4.3-3 Spearman rank correlation coefficients for the basic NI and modified Nutrition Indices

	NI based on unstandardised components			
	Full Sample *(n=106)*	Low NI Group *(n=35)*	Medium NI Group *(n=35)*	High NI Group *(n=36)*
NI, components standardised (2.1)				
Rank Correlation Coefficient	0.994 ***	0.966 ***	0.916 ***	0.959 ***
P-Value	(0.000)	(0.000)	(0.000)	(0.000)
NI, components standardised (2.2)				
Rank Correlation Coefficient	0.995 ***	0.973 ***	0.927 ***	0.965 ***
P-Value	(0.000)	(0.000)	(0.000)	(0.000)
NI, components standardised (2.3)				
Rank Correlation Coefficient	0.992 ***	0.958 ***	0.899 ***	0.935 ***
P-Value	(0.000)	(0.000)	(0.000)	(0.000)
NI with modified aggregation				
Rank Correlation Coefficient	0.994 ***	0.971 ***	0.943 ***	0.952 ***
P-Value	(0.000)	(0.000)	(0.000)	(0.000)

Source author's own calculation

Note The standardisation of NI components is carried out according to Formulas (2.1), (2.2) and (2.3). The aggregation function is modified by using the weighted average of second order in Formula (4.1), see Chapter 2.3.3 for further explanation.

For the total sample, rank correlation coefficients are close to unity. The coefficients are still above 0.90 (the upper threshold of redundancy proposed in Chapter 2.1.2) for the subgroups, except for the correlation of the NI based on normalised indicators with medium values of the basic NI, for which the coefficient amounts to 0.899. Given the disadvantages of standardisation discussed in Chapter 2.3.2.2 – the arbitrariness of limit setting as well as the problems with intertemporal comparability and ready interpretation of index values – the high degree of redundancy of the four different index versions is a strong argument against standardisation. There is no rationale to apply standardisation procedures if this results in insignificant changes in the country ranking compared to a ranking according to an index based on unstandardised components, as is the case here. The same holds for the variation of the aggregation function by subtracting a weighted average of second order from 100 for NI calculation (instead of the weighted average of first order as in Formula (4.1)[26]. Though this type of aggregation might be conceptually appealing because it implies restricted substitutability of partial indicators, the resulting NI values are again highly redundant vis-à-vis the basic version of the NI (see the last row of Table 4.3-3).

Table 4.3-4 Spearman rank correlation coefficients for the basic NI (with weights derived from factor analysis) and Nutrition Indices with varied weights

	weights of modified NI			NI	Low NI	Medium NI	High NI
	PUN	CUW	CM	*(n=106)*	*(n=35)*	*(n=35)*	*(n=36)*
NI 11	***0.10***	0.45	0.45	0.965 ***	0.780 ***	0.768 ***	0.761 ***
NI 12	0.45	***0.10***	0.45	0.980 ***	0.889 ***	0.776 ***	0.863 ***
NI 13	0.45	0.45	***0.10***	0.995 ***	0.960 ***	0.934 ***	0.979 ***
NI 21	***0.20***	0.40	0.40	0.989 ***	0.930 ***	0.899 ***	0.912 ***
NI 22	0.40	***0.20***	0.40	0.993 ***	0.956 ***	0.914 ***	0.946 ***
NI 23	0.40	0.40	***0.20***	0.997 ***	0.977 ***	0.969 ***	0.989 ***
NI 41	***0.40***	0.30	0.30	0.998 ***	0.990 ***	0.975 ***	0.982 ***
NI 42	0.30	***0.40***	0.30	0.998 ***	0.977 ***	0.979 ***	0.982 ***
NI 43	0.30	0.30	***0.40***	0.999 ***	0.995 ***	0.988 ***	0.995 ***
NI 61	***0.60***	0.20	0.20	0.978 ***	0.909 ***	0.713 ***	0.898 ***
NI 62	0.20	***0.60***	0.20	0.973 ***	0.781 ***	0.827 ***	0.866 ***
NI 63	0.20	0.20	***0.60***	0.987 ***	0.906 ***	0.851 ***	0.918 ***

Source author's own calculation

Note All coefficients are significant at the 0.001 level.

Changing the weights is another option to modify the aggregation function. Additional NI versions with varied weights are computed to check if the NI provides scope for 'weight pluralism' (see Chapter 2.3.3). Spearman rank correlation coefficients for the basic NI and indices based on varied weighting factors are shown in Table 4.3-4. Unless the weights are changed considerably – raising the weight of one component to 0.60 or reducing it to 0.10 whilst assigning half the difference to unity to each of the other two components - the modified NI versions are still highly redundant for the subsamples as compared to the basic NI according to Formula (4.1). The correlation coefficients for the total sample range from 0.965 to 0.985 for more extreme variations of weights, and are higher for moderate changes between –43% and +25% (see the middle rows of Table 4.3-4). These findings suggest that although the weighting of

[26] The weighted average of second order implies the weighting of squared partial indicators and taking the square root of their sum (compare Formula (2.4) in Chapter 2.3.3).

composite indices is always criticised for its arbitrariness, it is not worthwhile discussing the weights extensively in the case of the NI. As stated for the HDI, it might be much more promising to invest time and resources into improving the underlying data.

As has been postulated by Ryten (2000) for composite indices, the basic NI is found to be quite robust in the face of changes in its aggregation function resulting either from different types of standardisation applied to its components, or the alternative use of the weighted average of second order, or moderate variations of weighting factors. For the sake of simplicity, unstandardised partial indicators and Formula (4.1)[27] are used for the final Nutrition Index values presented and discussed in the subsequent chapter.

[27] The use of almost equal weighting factors for underweight in children and the under-five mortality rate may raise some concern. A scenario could be imagined where more children survived, but all in a malnourished state. This shift is of course an improvement, but would the NI be able to reflect this if the percentage of children dying before age five and the percentage of children underweight are assigned nearly equal weights? The answer is yes, since the mortality rate is expressed as a percentage of the total cohort of children born at a given point in time, whereas child malnutrition is referenced to all children under five years. Thus, for the above scenario, the under-five mortality rate would inevitably decrease more than the percentage of underweight children increased, thereby inducing a rise in the NI value.

5 Nutrition Index: Ranking and Trends

NI scores are used in the following chapter to set up a ranking of countries and to illustrate the world nutrition situation in a map. Furthermore, they are found to be helpful in showing trends in food security and nutrition for countries and regions.

5.1 Ranking and Mapping of Countries

The ranking of countries according to their Nutrition Index is shown in Table 5.1-1, with the best performers at the top of the list[1]. The international ranking of 106 developing countries and countries in transition covers a range of about 43 NI points. Kuwait is at the top of the list with an NI of 98.0 and is closely followed by Yugoslavia (Serbia and Montenegro), Chile, Libya, Argentina and Lebanon. Countries that experienced long-lasting violent conflicts affecting the infrastructure and productive base of the economy have very low NI scores, indicating a miserable track record in food security and nutrition. Nine out of the eleven countries at the very bottom of the list – Afghanistan, Ethiopia, Somalia, Eritrea, Mozambique, the Democratic Republic of Congo, Angola, Chad and Sierra Leone - were torn by civil wars in the NI reference period, or have only recently started to recover from cruel internal conflicts. With an NI of only 54.8, Afghanistan is by far the worst performer in the international comparison of Nutrition Index values, which is largely the outcome of 18 years of uninterrupted warfare until 1997 (HIIK 2000). Niger belongs in the list of the bottom five countries although it was not affected by war. Yet a violent crisis took place from 1990 to 1995 (HIIK 2000), and the country is part of the ecologically vulnerable Sahel zone with irregular rainfalls. (The case of Bangladesh, which is among the bottom ten countries despite being neither afflicted by war nor by a violent crisis throughout the 1990s, is discussed in Chapter 6.3.1.) Warfare was frequently accompanied by economic mismanagement such as undue price controls, barriers to internal trade and market development, that were set up by the state. A few examples of war-torn countries with such mistaken economic policies are Angola, Ethiopia (formerly including Eritrea) and Mozambique (Reichel 1991, Azam et al. 1994).

---

[1] Different NI values and ranking positions as compared to the results published in Wiesmann et al. 2000 are due to updated NI values (using recently published data from WHO 2000a). The underlying data are shown together with respective NI values in Table A 5.1-1 in the Appendix. Changes in the analyses relating to GNP per capita in Chapter 6.3.1 have also been effected by revised PPP estimates (World Bank 2000).

Table 5.1-1 Nutrition Index (NI) – ranking of countries

NI Rank	Country	Nutrition Index 1981	1992	1997	NI Rank	Country	Nutrition Index 1981	1992	1997	NI Rank	Country	Nutrition Index 1981	1992	1997
1	Kuwait	94.2	89.0	**98.01**	**37**	Iraq*	91.1	**90.74**	..	**73**	Congo, Rep.*	73.5	**77.98**	..
2	Yugoslavia[1]*	..	..	**97.74**	**38**	Swaziland	..	**90.47**	..	**74**	Uganda*	75.5	78.6	**77.81**
3	Chile	96.1	96.1	**97.64**	**39**	Ecuador	..	**90.09**	..	**75**	Pakistan	68.4	75.7	**76.75**
4	Libya	93.4	94.9	**97.29**	**40**	Armenia*	..	..	**89.70**	**76**	Kenya	80.8	77.1	**76.35**
5	Argentina	97.1	98.0	**97.10**	**41**	Peru	80.9	81.2	**89.27**	**77**	India	59.5	67.6	**74.43**
6	Lebanon	91.5	95.1	**97.09**	**42**	Dominic. Rep.*	84.0	85.8	**87.68**	**78**	Nepal	57.4	72.4	**74.42**
7	Jordan	.	95.7	**96.54**	**43**	Uzbekistan	..	..	**87.10**	**79**	Mali	58.7	74.5	**74.31**
8	Uruguay	95.5	94.6	**96.54**	**44**	Kyrgyz Rep.	..	..	**87.08**	**80**	Burkina Faso	60.0	74.6	**73.72**
9	Russian Federation	..	..	**96.45**	**45**	Bolivia	81.3	83.9	**86.62**	**81**	Rwanda*	..	72.5	**73.10**
10	Romania	..	**96.08**	..	**46**	Azerbaijan*	..	..	**86.35**	**82**	Central Afric. Rep.	68.8	68.7	**72.66**
11	Cuba	95.4	**95.97**	..	**47**	Gabun	83.8	**86.15**	..	**83**	Guinea	73.0	**72.20**	..
12	Costa Rica	94.4	96.7	**95.88**	**48**	Guyana*	.	84.1	**85.99**	**84**	Laos	70.8	74.0	**72.04**
13	Tunisia	91.0	95.1	**95.64**	**49**	Thailand	77.1	82.8	**84.80**	**85**	Tanzania	77.7	75.0	**72.00**
14	Turkey	90.2	93.0	**95.11**	**50**	Indonesia*	72.3	80.0	**84.76**	**86**	Comoros	..	70.7	**71.38**
15	Kazakhstan	..	..	**94.88**	**51**	Myanmar[2]	75.2	82.8	**84.68**	**87**	Malawi	74.5	68.5	**70.70**
16	Jamaica	93.0	94.1	**94.64**	**52**	Botswana	76.5	83.2	**84.53**	**88**	Zambia	78.3	72.0	**70.56**
17	Fiji	..	..	**94.62**	**53**	Guatemala*	75.7	83.7	**83.95**	**89**	Liberia*	77.9	**69.80**	..
18	Syria	91.3	94.0	**94.39**	**54**	Ghana	64.5	75.8	**83.89**	**90**	Yemen	61.6	73.4	**69.54**
19	Croatia	..	..	**94.01**	**55**	Nicaragua*	83.2	..	**83.87**	**91**	Burundi*	72.4	**69.33**	..
20	South Africa	..	..	**93.74**	**56**	Honduras*	79.5	84.7	**83.37**	**92**	Madagascar	76.9	70.0	**68.80**
21	Saudi-Arabia	91.1	**93.65**	..	**57**	Côte d'Ivoire	86.9	87.2	**83.00**	**93**	Djibouti	..	**68.24**	..
22	Brazil	89.6	91.3	**93.34**	**58**	Senegal	79.8	82.6	**82.91**	**94**	Cambodia*	53.8	68.1	**68.10**
23	Iran	78.8	.	**93.31**	**59**	Mongolia	..	**82.11**	..	**95**	Haiti*	65.7	66.1	**66.45**
24	Morocco	86.3	**93.25**	..	**60**	Mauritania	69.8	72.7	**81.95**	**96**	Sierra Leone*	68.8	**66.36**	..
25	Trinidad & Tobago	93.8	**93.16**	..	**61**	Philippines	78.0	80.3	**81.83**	**97**	Bangladesh*	56.3	64.7	**65.63**
26	Algeria*	86.2	93.0	**92.89**	**62**	Togo	76.2	77.8	**80.94**	**98**	Chad*	58.1	63.9	**65.47**
27	Egypt	86.3	93.3	**92.70**	**63**	Lesotho	81.1	79.3	**80.85**	**99**	Angola*	72.9	62.0	**63.99**
28	Mexico*	90.2	**92.69**	..	**64**	The Gambia	..	81.4	**80.32**	**100**	Congo, Dem. Rep.[3]*	71.7	71.0	**63.60**
29	Malaysia	87.7	90.2	**92.52**	**65**	Namibia	..	**80.27**	..	**101**	Mozambique	58.7	53.4	**63.58**
30	Mauritius	86.3	91.8	**92.45**	**66**	Benin	71.3	80.5	**79.86**	**102**	Niger	62.6	61.6	**60.06**
31	Colombia*	88.6	90.5	**92.30**	**67**	Nigeria	70.2	77.7	**79.62**	**103**	Eritrea*	..	..	**59.85**
32	Paraguay	91.3	**91.69**	..	**68**	Vietnam	68.4	75.7	**79.49**	**104**	Somalia*	59.8	**57.10**	..
33	El Salvador*	83.5	89.1	**91.66**	**69**	Cameroon	81.8	81.3	**79.40**	**105**	Ethiopia	61.0	**56.45**	..
34	Venezuela	93.9	93.9	**91.58**	**70**	Zimbabwe	78.2	80.1	**79.37**	**106**	Afghanistan*	72.6	59.3	**54.77**
35	Panama	86.6	**91.41**	..	**71**	Sri Lanka*	75.7	77.8	**79.02**					
36	China*	80.2	87.3	**91.19**	**72**	Sudan*	76.6	74.5	**78.53**					

Notes Countries for which data availability did not permit NI calculation, but where nutrition problems probably exist: Albania, Bahrain, Belarus, Bhutan, Bosnia & Herzegovina, Bulgaria, Estonia, Georgia*, Guinea-Bissau*, North Korea*, Latvia, Lithuania, Macedonia, Moldavia, Oman, Papua New Guinea (NI 1981: 76,9), Qatar, Slovak Rep., Tadjikistan, Turkmenistan*, Ukraine

1 Serbia and Montenegro 2 former Burma 3 former Zaire

*countries where the nutrition situation has probably been affected by armed conflicts or natural disasters since 1997

Source author's own calculations based on data from FAO 1999a and 1999b, UN ACC/SCN 1993, WHO 2000a, UNICEF 1995 and 1999

Figure 5.1-1 Nutrition Index (NI) – mapping of countries

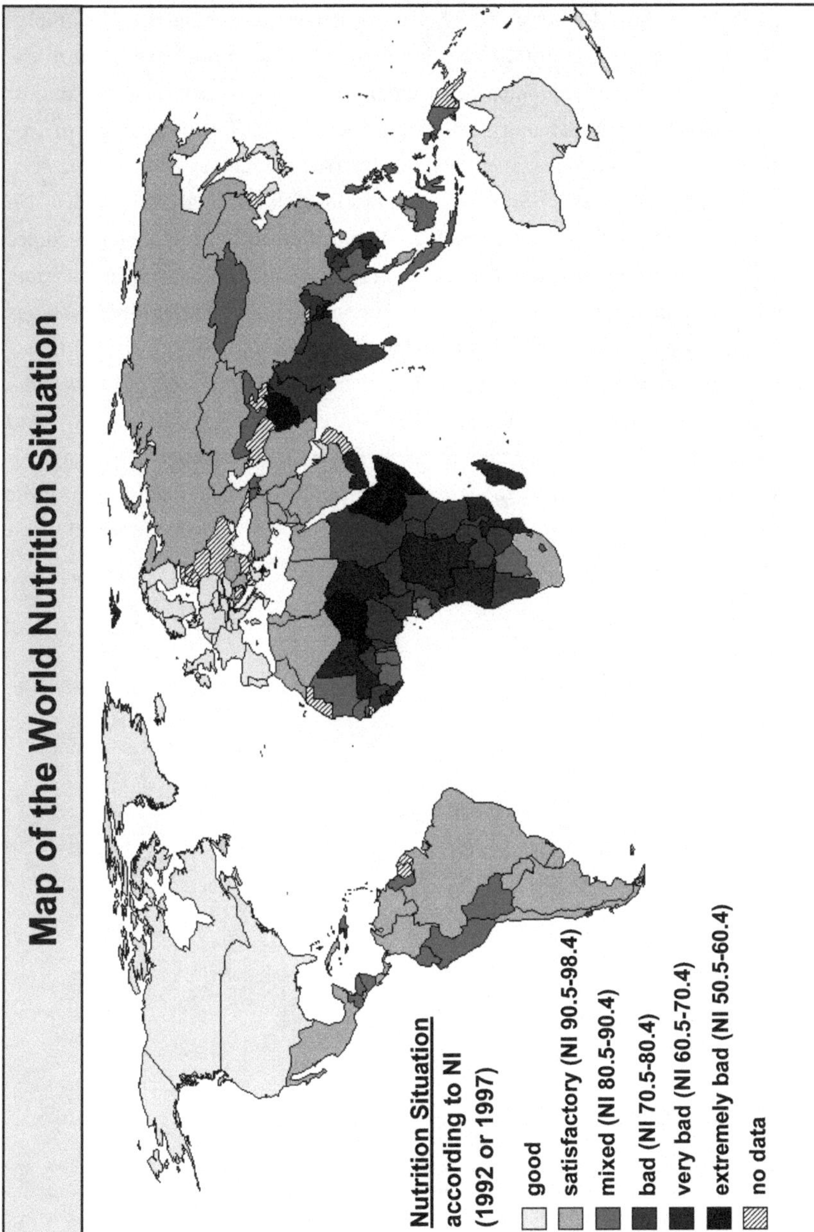

Source author's own presentation

In the world map depicted in Figure 5.1-1, countries with NI 1997 are represented (NI 1992 for those for which NI 1997 could not yet be computed). The graphical illustration of the world nutrition situation shows that countries with a bad to extremely bad nutrition situation, indicated by an NI value of less than 80.4[2], are almost exclusively located in Sub-Saharan Africa and South Asia. The connection of low economic performance and a desolate nutrition situation, which has already been seen in the ranking, is also evident from the concentration of severe nutrition problems in very poor Sub-Saharan African countries according to the world map. Rising incomes are likely to lead to an improved nutrition situation, but underinvestment in nutrition, on the other hand, may keep productivity low[3]. The ranking of countries within regions is discussed in the next section, together with the development of NI values over time.

The results of calculating NI values for the states and provinces of the world's two most populous countries – India and China – are shown in the Boxes 5.1-1 and 5.1-2 on the following pages. The findings underscore the importance of subnational disaggregation: vast disparities in food security and nutrition are found within these countries, expressed in differences of more than ten NI points between best and worst performing administrative regions.

[2] The mean NI of countries in this group amounts to 71, corresponding to average values of 39% undernourished (which is roughly equivalent to a dietary energy supply per capita of 2000 kcal per capita per day according to the model devised in section 4.1.1), 34% underweight children and an under-five mortality rate of 160 per 1000 live births.

[3] The relationship between GNP per capita and NI scores is more closely investigated in Chapter 6.3.1.

Box 5.1-1 Nutrition Index values for Chinese Provinces

In Table B 5.1-1-1, Chinese Provinces are ranked according to their NI 1998 values which are shown together with their underlying data.

Table B 5.1-1-1 Chinese Provinces ranked by NI 1998 and underlying data

	undernourished in population[1] (in %) 1998	underweight children under five years of age[2] (in%) 1998	under-five mortality[2] (per 1000 live births) 1998	Nutrition Index (NI) 1998
1 Shanghai	3.7	0.3	8	98.41
2 Jiangsu	5.2	1.7	18	97.13
3 Beijing	7.7	0.8	10	96.86
4 Tianjin	7.9	1.0	13	96.61
5 Zhejiang	6.4	1.6	22	96.60
6 Liaoning	5.8	2.3	23	96.55
7 Guangdong	4.9	2.7	34	96.34
8 Chongqing	6.9	1.6	29	96.20
9 Fujian	5.4	3.1	31	96.14
10 Jiangxi	2.1	5.5	47	95.87
11 Jilin	7.8	2.9	23	95.71
12 Inner Mongolia	5.7	3.4	41	95.60
13 Heilongjiang	4.2	8.1	20	95.31
14 Hubei	2.5	10.0	32	94.84
15 Hunan	1.2	13.8 a	33 a	94.03
16 Shandong	13.6	2.0	30	93.86
17 Qinghai	8.1	6.0	55	93.48
18 Anhui	11.1	4.1	46	93.39
19 Guangxi	13.0	5.0	30	93.07
20 Hainan	13.4	4.8 a	42	92.58
21 Gansu	12.6	5.1	54	92.35
22 Henan	16.8	6.4	28	91.43
23 Shanxi	19.9	3.7	26	91.35
24 Tibet	17.5	3.5	73	90.55
25 Ningxia	8.0	17.6 a	43	90.21
26 Guizhou	14.7	8.9	71	89.80
27 Sichuan	7.5	18.7	59	89.48
28 Yunnan	10.2	17.0	51	89.41
29 Shaanxi	19.0	11.2	37	88.88
30 Xinjiang	8.3	20.0 a	74	88.27
31 Hebei	25.7 b	11.0	30	86.94
China total	9.8 *	7.3 **	37.54 ***	93.13 ****

Source own calculations and data from UNICEF (personal communication)

Notes 1 estimated from food consumption data, author's own calculation

2 personal communication from Mark Hereward from UNICEF

* 13% in 1995-97 and 11% in 1996-98 according to FAO (FAO 1999a and 2000a)

** 9.0% in 1998 according to survey data (WHO 2000a)

*** 42 per thousand in 1998 according to the Child Mortality Surveillance System of the Ministry of Health (figure obtained from Mark Hereward from UNICEF)

**** NI 1997 for China: 91.2

a data are of 1995

b doubts about reliability of estimate

Box 5.1-1 (continued) Nutrition Index values for Chinese Provinces

The aggregated NI for China as a whole amounts to 93.1. This value is higher than the NI calculated for 1997 from nationally representative data published by FAO 1999a, WHO 2000a and UNICEF 1999 (which has a value of 91.2). Although the difference can be partly attributed to the positive overall trend in China's nutrition situation, another reason may be that the figures aggregated from the provincial data represent slight underestimates, which is one result of imperfections in the provincial data and estimates. However, the disparities in NI values among provinces are much larger than the deviation of the NI 1998 from NI 1997, which underscores the necessity for subnational disaggregation. The NI value of Shanghai – the province with the highest ranking - exceeds Hebei's NI by 11.5 points. Even if the probable overestimate of undernourishment in Hebei is taken into account and the NI values of Shanghai (NI 98.4) and Xinjiang (NI 88.3) are compared instead, the difference is still larger than 10. According to NI 1998, the two bottom provinces in the ranking are doing hardly better than Bolivia, which had an NI of 86.6 in 1997, whereas the nutrition situation in Shanghai is even better than in Kuwait, which led the international ranking of 1997 with an NI of 98.0.

In Figure B 5.1-1-1, NI values for Chinese Provinces are mapped, and Figure B 5.1-1-2 shows income by province for 1998.

Figure B 5.1-1-1 Map of the nutrition situation in China according to NI 1998

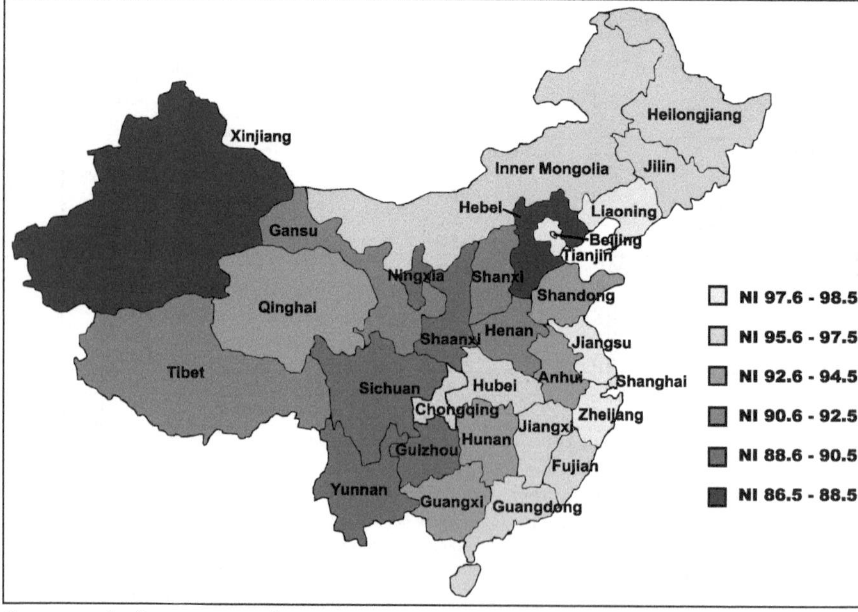

Source author's own presentation

Box 5.1-1 (continued) Nutrition Index values for Chinese Provinces

The maps suggest an association between low provincial incomes and a relatively bad nutrition situation. The poor rural provinces in Central and Western China have relatively low NI values. This also holds for Hebei with a net income of 2988 Yuan per capita per year, though it is suspected that its NI is underestimated. In general, the coastal areas are in a favourable position both with respect to incomes and food security and nutrition. A more detailed analysis of the relationship between net incomes and NI values of China's Provinces is undertaken in Chapter 6.3.

Figure B 5.1-1-2 Map of net incomes in China in 1998

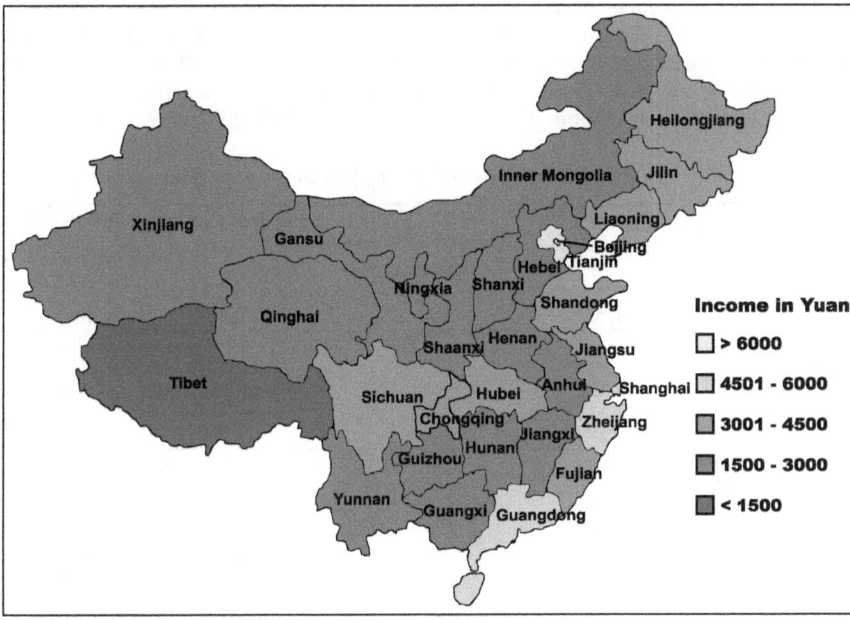

Source author's own presentation, based on data from National Bureau of Statistics of China 1999

Box 5.1-2 Nutrition Index values for Indian Federal States

The NI 1994 values for major Indian States are listed in Table B 5.1-2-1 together with the underlying data. The NI 1994 value for India as a whole, aggregated from the values for the 15 states, is 70.3. This is in line with the trend marked by the NI values for India calculated for the years 1992 and 1997, which are 67.6 and 74.4, respectively. The importance of a subnationally disaggregated view is confirmed by the data in Table B 5.1-2-1. Both the NI values and the underlying indicators show wide disparities between states. The difference between the highest NI value for Haryana (NI 79.7), and the lowest value for Bihar (NI 66.5) amounts to more than 13. According to the Nutrition Index, the nutrition situation in 1994 in Bihar was no better than that in Haiti in 1997, which arrived at the same NI value at that time, whereas the NI for Haryana came close to that for Benin. This West African state had made considerable progress in staple food production within the last decade and achieved an NI of 79.9 in 1997. However, the NI values of all the Indian States are below 80.4, thus indicating a bad to extremely bad nutrition situation (compare the NI categories listed in Figure 5.1-1).

Table B 5.1-2-1 Indian States ranked by NI 1994 and underlying data

		undernourished in population[1] (in %) 1993-94	underweight children under four years of age[2] (in%) 1992-93	under-five mortality[2] (per 1000 live births) 1992-93	Nutrition Index (NI) 1994
1	Haryana	13.6	37.9	98.7	79.7
2	Kerala	35.9	28.5	32.0	78.0
3	Rajasthan	15.9	41.6	102.6	77.6
4	Punjab	16.8	45.9	68.0	77.1
5	Orissa	21.9	53.3	131.0	70.8
6	Andhra Pradesh	31.3	49.1	91.2	70.6
7	West Bengal	22.8	56.8	99.3	70.5
8	Uttar Pradesh	19.1	59.0	141.3	69.5
9	Karnataka	30.0	54.3	87.3	69.4
10	Gujarat	33.3	50.1	104.0	69.2
11	Madhya Pradesh	25.1	57.4	130.3	68.5
12	Tamil Nadu	40.7	48.2	86.5	68.0
13	Maharashtra	36.4	54.2	70.4	68.0
14	Assam	33.8	50.4	142.2	67.6
15	Bihar	26.2	62.6	127.5	66.5
	India total	25.8 *	53.2	109.3	70.3 **

Source own calculations and data from International Institute for Population Sciences 1995

Notes 1 estimated from dietary energy supply data, author's own calculation (see Chapter 4.1.3)
2 International Institute for Population Sciences 1995

* 26% according to FAO for 1990-92

** NI for India in 1992 and 1997: 67.6 and 74.4, respectively

Box 5.1-2 (continued) Nutrition Index values for Indian Federal States

In Figure B 5.1-2-1, NI values for the major Indian States are illustrated. A simultaneous look at the main cropping areas of the two most important Indian staple foods, rice and wheat, provides insight (see Figure B 5.1-2-2). In general, food crop productivity can be expected to be high in regions where the largest part of the main staples is grown. In two out of the three major wheat surplus producing states – Haryana and Punjab - the nutrition situation is in fact quite favourable compared to other Indian states. Rajasthan, the third Northern Indian State with a relatively high NI value, covers at least a part of the wheat production area.

Figure B 5.1-2-1 Map of the nutrition situation in India according to NI 1994

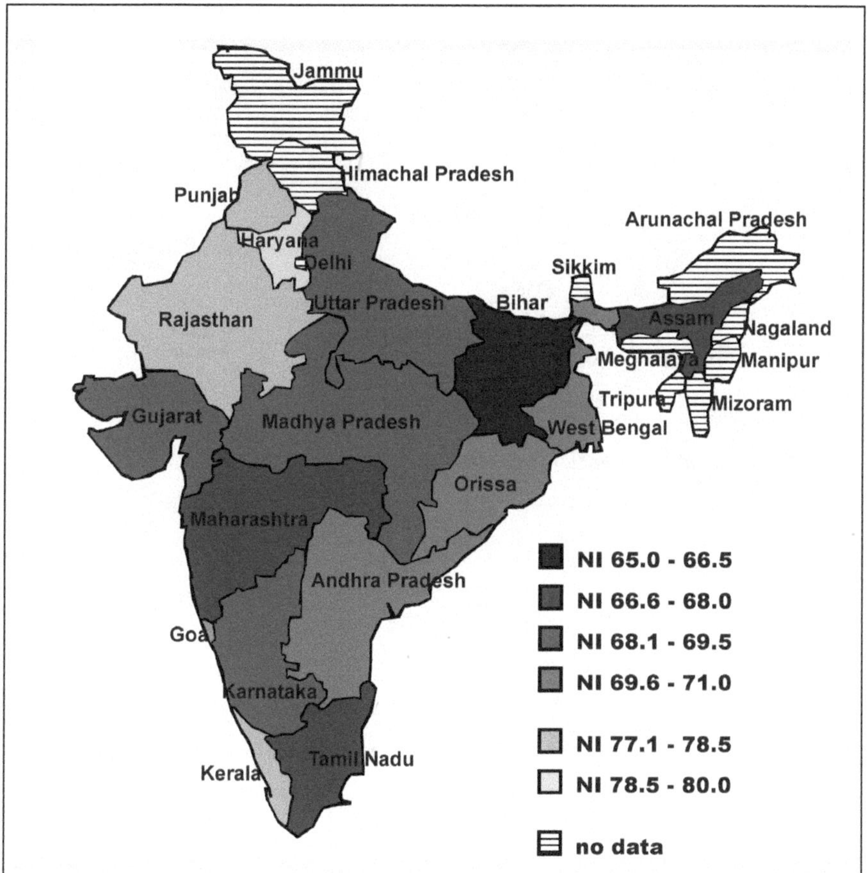

Source author's own presentation

Box 5.1-2 (continued) Nutrition Index values for Indian Federal States

In contrast, the neighbouring state of Uttar Pradesh, one of the wheat surplus producers, failed to reach a comparatively high NI, the value being only 69.5 in 1994. The main rice growing states along the Eastern coast – West Bengal, Andhra Pradesh and Orissa – were doing better than most other Indian States in terms of the NI, though not as well as the major wheat producers Haryana and Punjab, and the state of Kerala with NI 78.0. Why Kerala reached such a high NI value in relation to other Indian States, despite being outside the agriculturally most productive regions, and reasons for Uttar Pradesh's weak performance (given its agricultural potential) are discussed with reference to economic and social indicators in Chapter 6.3.

Figure B 5.1-2-2 Map of main cropping areas of rice and wheat in India

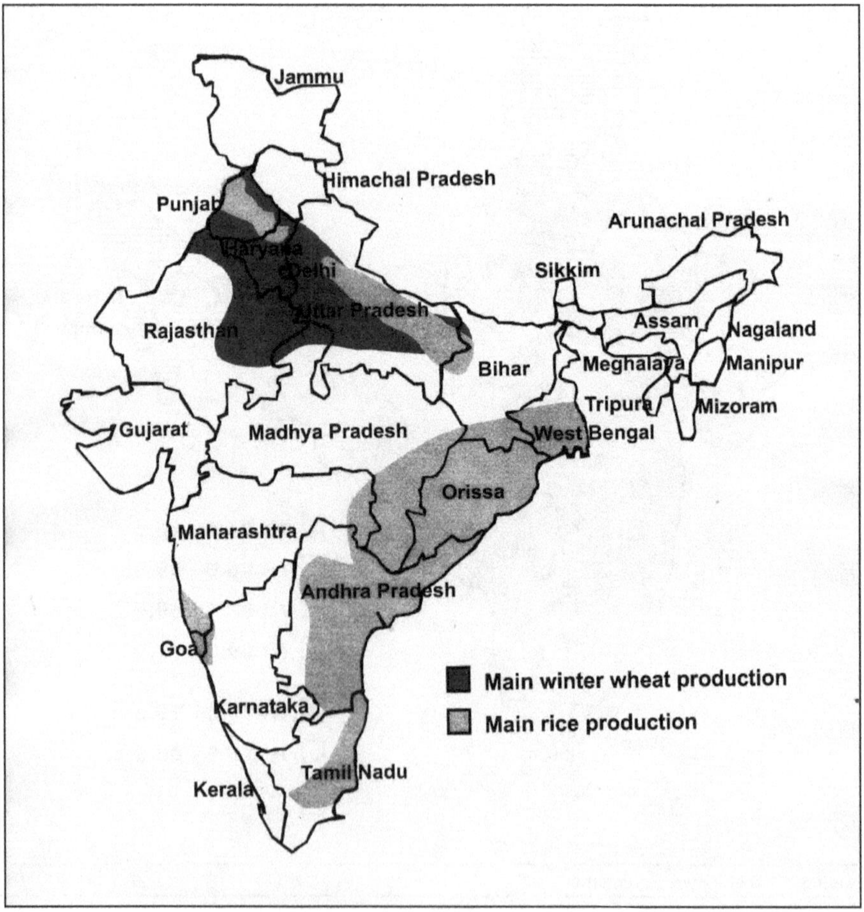

Source adapted from FAO/GIEWS 2001

5.2 Regional Comparisons and Trends

Before looking in detail at the development of the nutrition situation in various regions, NI trends for aggregates of three continents[4] are presented in Figure 5.2-1 (see next page). In **Africa** as a whole - including North Africa - little progress has been made since the beginning of the 1980s, and stagnation is found throughout the 1990s at an NI value of 77.0. Slight declines in the percentage of undernourished and the under-five mortality rate were outweighed by rising underweight prevalence in children (see Table 5.2-1).

In contrast to the lack of development in Africa, **Asia** experienced considerable improvements in its overall nutrition situation. Starting with the lowest regional NI of 72.2 in the 1980s, Asia surpassed the African continent by 1992 due to notable reductions in undernourishment, underweight prevalence and child mortality, leading to a gain of eight points in NI value by 1997. One of the reasons for this is that the introduction of high-yielding rice and wheat varieties together with irrigation, fertiliser and pesticides in the 1960s – the so-called Green Revolution - has been far more successful in Asia than in Africa, where this technology package has not been widely applied (von Braun 1996). Cereal yields increased by 160% in Asia from 1961 to 1997, but by only 50% from a lower initial level in Africa within the same period (FAO 1999b). Despite its heavily criticised negative side-effects, the Green Revolution has had a positive impact: rapid agricultural output growth in Asia fuelled economic growth and public investment in rural areas, benefiting food security and nutrition (von Braun 1996).

However, the NI 1997 for Asia is still far from the NI value of 89.3 reached in **Latin America and the Caribbean** in 1981. Ongoing progress can be seen in this region, though not at a great pace, with slowly falling levels of undernourishment, underweight prevalence and more pronounced reductions in child mortality since 1980. From 1981 to 1997, the NI increased by about three points (see Table 5.2-1 on the following page).

---

[4] Due to limited NI coverage, the percentage of undernourished and the under-five mortality rate were weighted with population size and aggregated separately for the UN regions (developed countries in Asia and the Pacific, like Japan, Australia and New Zealand, and the successor states of the former Soviet Union were not taken into account). In the case of underweight prevalence, estimates for 1980 (UN ACC/SCN 1993) were aggregated in the same manner, and the figures published by WHO for the years 1990 and 1995 for the main developing regions were adopted directly (WHO 1997). The NI was computed subsequently from the results for the partial indicators.

Figure 5.2-1 NI trends 1981-1997 in regional aggregates

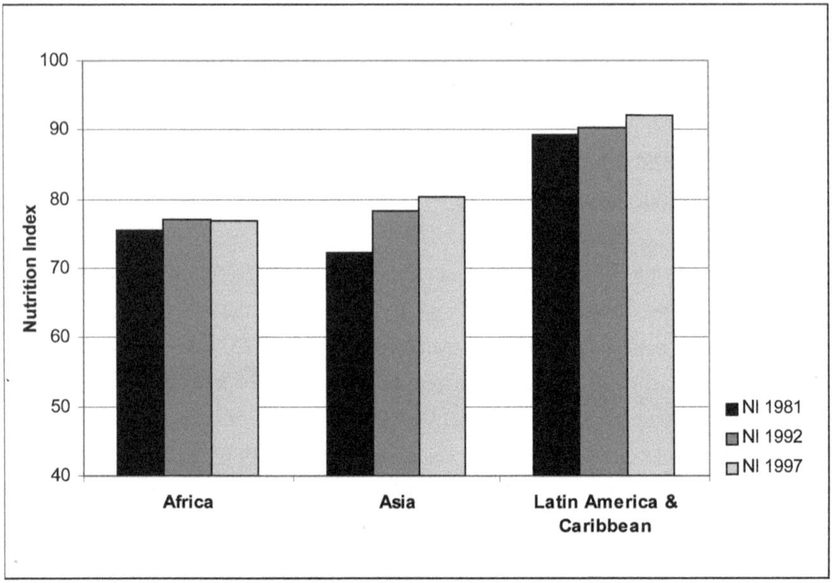

Source author's own presentation based on data presented in Table 5.2-1

Table 5.2-1 Nutrition Index and underlying data, aggregated by region

	Undernourished [1]			Underweight [2]			Under-five mortality [3]			Nutrition Index		
	1979-1981	1990-1992	1995-1997	1980	1990	1995	1980	1992	1997	1981	1992	1997
Africa	29.3	27.8	26.3	24.7	26.0	28.4	193	151	144	75.6	77.0	77.0
Asia	30.5	19.8	16.7	41.2	37.2	35.0	115	81	73	72.2	78.3	80.3
Latin America & Caribbean	12.2	13.1	10.8	11.6	11.4	9.5	82	46	35	89.3	90.3	92.1

Source author's own calculations based on data from FAO 1999a and 1999b, UN ACC/SCN 1993, WHO 1997, UNICEF 1995 and 1999

Notes 1 in the population (in %)
2 in children under five (in %)
3 per 1000 live births

Clearly, highly aggregated NI scores can easily conceal disparities among and within subregions. The first overview given here is further differentiated in Figure 5.2-2 on page 90f, where states are grouped by regions and ranked by their NI 1997 (only countries where data availability allowed the calculation of NI 1997 are shown).

In **North Africa and the Middle East**, improvements in the nutrition situation dominate the picture during the 1981-1997 period. NI values are above 90.5 in 1997, indicating a satisfactory nutrition situation for the countries in this region, except for Yemen which had an NI of only 69.5 in 1997.

In contrast to these high scores, most **West and Central African** states are characterised by low NI values (below 81), reflecting a bad nutrition situation. But fortunately, the majority of countries in this region have experienced positive trends since the beginning of the 1980s. The largest rise was seen in Ghana, where the NI jumped from 64.5 in 1980 to 83.9 in 1997 - a gain of almost 20 points, which is unique for the entire set of countries under consideration. The percentage of undernourished was reduced from 61% in 1979-81 to 11% in 1995-97. Along with the introduction of improved yams, maize, rice and cassava varieties - the yields of this main staple rose by almost 40% from 1980 to 1996 - and an increase of the cropped area by 26%, rapid economic growth in other sectors helped to boost food supply (FAO 1999a). In the years 1980 to 1983, Ghana experienced negative GNP per capita growth, but this decline was reversed from 1984 onwards after an Economic Recovery Programme (a structural adjustment programme) had been implemented starting in 1983, which resulted in new investment and rising exports (World Bank 2000, Berry 1994). Marked improvements in access to sanitation, health and education were also achieved during that period (FAO 1999a). Accordingly, the under-five mortality rate fell from 157 per thousand live births in 1980 to 107 in 1997. However, no marked decrease of underweight in children was observed - it still amounted to 27.3% in 1993-94 (see Table A 5.1-1). The dominant role of cassava and other starchy roots of similar nutritive value (providing around 50% of dietary energy in the Ghanaian diet, calculated from FAO 1999b) is believed to be unfavourable for child growth and development[5]. Considerable, but smaller than in Ghana, improvements in food security and nutrition were found in Burkina Faso and Mauritania. The nutrition situation has worsened in Niger, Cameroon, Côte d'Ivoire and Sierra Leone since 1981. In civil-war-shaken Liberia, the downward trend of the NI is the most pronounced among all countries in this region, with a loss of 8.1 points (1992 as compared to 1981, no data for 1997, compare Table 5.1-1).

[5] In fact, cassava roots, the preferred staple in Ghana, contain little but carbohydrate – they have much less protein, B vitamins and iron than cereals. Protein deficiency is much more common in young children weaned on to cassava than in those weaned on to millet or maize (Latham 1997).

Figure 5.2-2 NI trends 1981-1997, countries grouped by regions

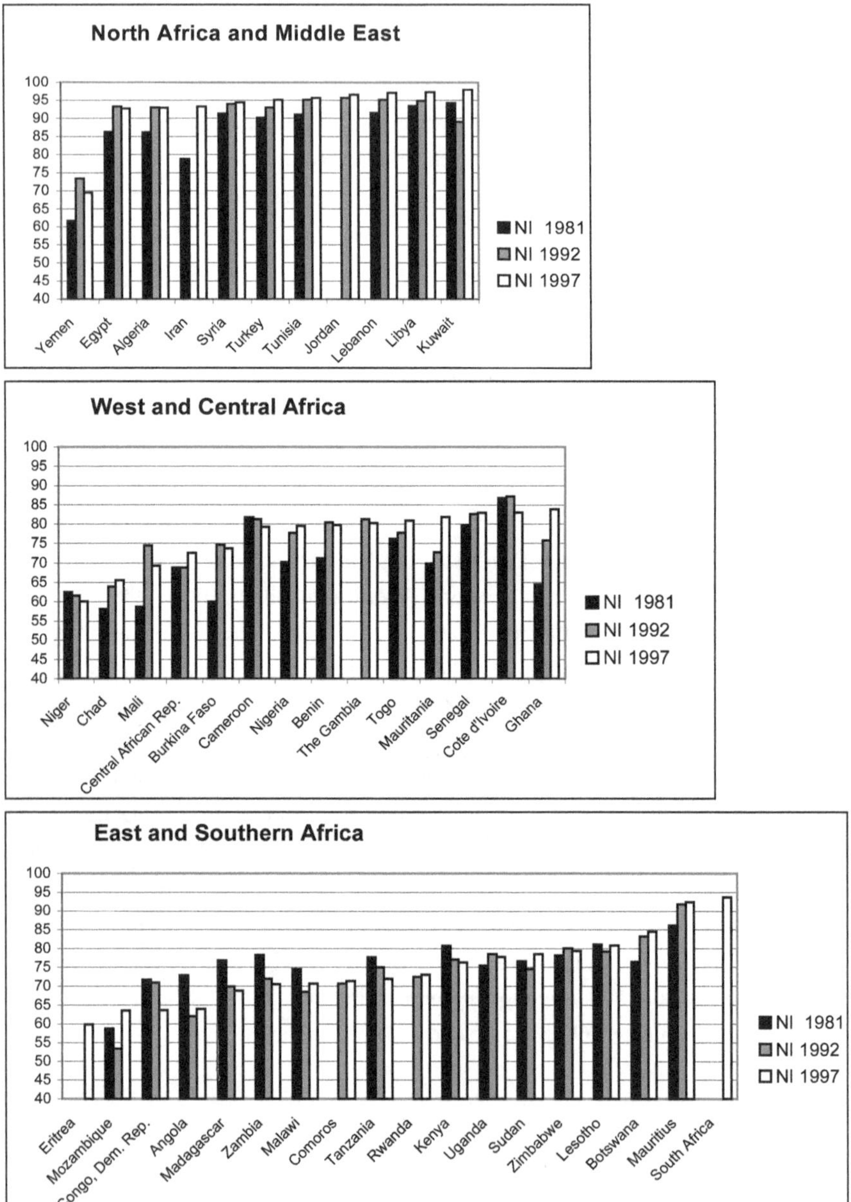

Source own presentation based on data shown in Table 5.1-1

Figure 5.2-2 (continued) NI trends 1981-1997, countries grouped by regions

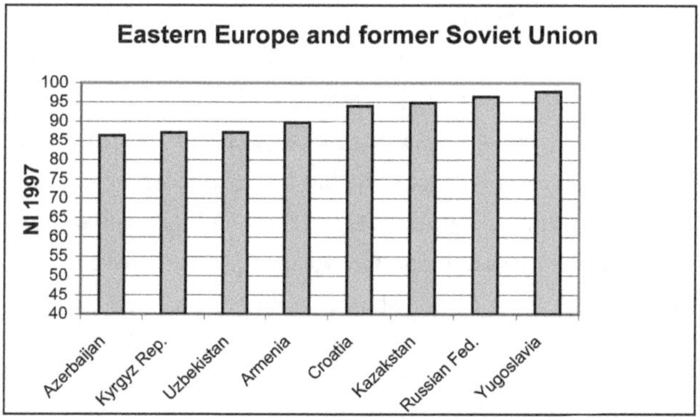

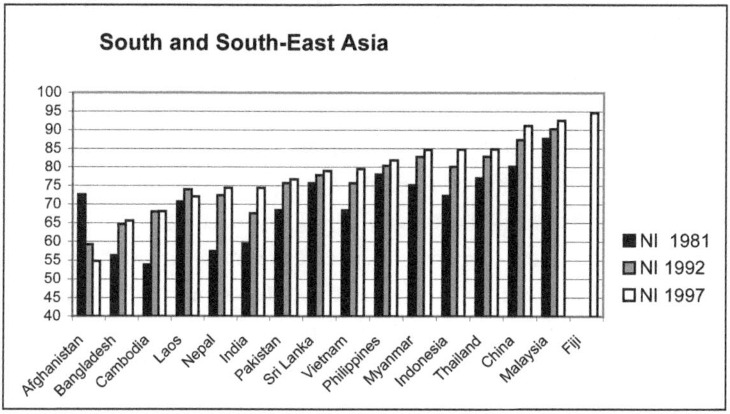

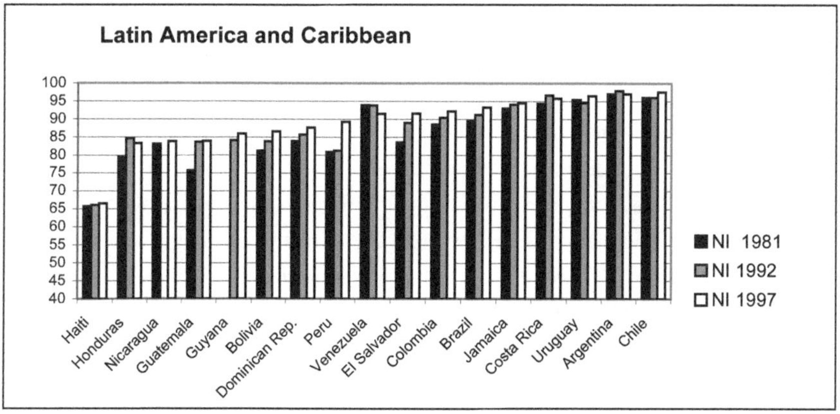

Source own presentation based on data shown in Table 5.1-1

In **Eastern and Southern Africa**, NI trends do not provide much scope for optimism. While Mozambique, Mauritius and Botswana show improvements in their nutrition situation since 1981, a great number of countries in this region suffer from a deterioration of food and nutrition security. This pertains especially to civil-war-shaken Angola and the war-torn Democratic Republic of Congo (former Zaire), where NI values decreased by 8.9 and 8.1 points respectively, but also to Madagascar, Zambia, Tanzania and Kenya. All three countries in this region for which NI scores could only be calculated for 1981 and 1992 – Burundi and the war-ravaged states of Ethiopia and Somalia - had NI scores below 73 in 1981 and showed a further decline during the following decade. The most serious downward trend could be observed for Ethiopia with a fall by 4.6 NI points between 1981 and 1992.

When looking at the NI values of the **South and South-East Asian** region in Figure 5.2-2, the dramatic fall of the NI for Afghanistan stands out. With a decrease by 17.9 points from 1981 to 1997, Afghanistan holds the worldwide record for the largest decline in NI values among all countries in the ranking. Several factors, most of them related to continuous warfare, contributed to the desperate situation: the population increased by 25% between 1980 and 1996, partly due to returning refugees. Cereal production fell slightly - more than 40% of arable land in Afghanistan is mined and cannot be farmed. The war-affected economy was unable to generate imports to fill the gap. As a consequence, dietary energy supply per capita fell significantly below requirements and the share of undernourished almost doubled from 33% in 1979-81 to 62% in 1995-97 (FAO 1999a). The percentage of underweight children was estimated to be 20.9% in 1980, and stood at 49.3% according to a survey in 1997 (see Table A 5.1-1 in the Appendix). Women, who are the main care-takers of children as a nutritionally vulnerable group, were deprived of their rights and opportunities by war, legislation and custom, particularly after the Taliban had seized power in 1996 (FAO 2002). In 1997, 50% of men and 81% of women were illiterate (World Bank 2000), and 2% of the population had been landmine casualties (FAO 1999a). Yet, apart from Afghanistan that is still facing a complex humanitarian emergency, the overall impression of NI trends in South and South-East Asia is positive due to improvements in the nutrition situation from 1981 to 1997 for all other countries in this region. Nepal, India, Cambodia, Indonesia, Vietnam and China experienced large rises in the NI of more than 10 points. Agricultural production in China and India rose markedly due to the Green Revolution. Cereal yields quadrupled in China and more than doubled in India between 1961 and 1997 – with an increase by 56% and 59%, respectively, from

1981 to 1997 (FAO 1999b) - and undernourishment declined considerably (see Table A 5.1-1 in the Appendix).

Moreover, China's and India's economies grew at impressive rates: in China, GNP per capita (in US$) increased fourfold from 1980 to 1997, and almost doubled in India during that period[6]. The percentage of the population with access to safe water was already high in China in the early 1980s (86%), but increased notably in India by the 1990s (from 54% to 81%) (World Bank 2000). Child malnutrition was reduced by nearly 15 percentage points in China and by more than 22 percentage points in India, whereas under-five mortality was cut back by around one third in both countries. But it should be stressed that despite the aforementioned notable progress in Vietnam, Cambodia, Nepal and India, the nutrition situation is still quite bad in these states and other countries in South Asia. Except for Myanmar, the NI values of these countries are below 80.4. At least, Malaysia's NI (92.5) reflects a satisfactory nutrition situation, and China has just crossed the threshold of 90.4 with an NI of 91.2 in 1997. The more recent effects of the Asian crisis in 1996/97 may not yet have been captured by the NI for 1997.

In **Latin America and the Caribbean**, the nutrition situation is fair to mixed in the vast majority of countries. On average, Latin American countries are better off than countries in the Caribbean, with NI values above 90.5 for six out of nine states. In some countries in the region that had reached high NI scores in 1981, only slight changes took place until 1997. Considerable progress is observed for Peru, Guatemala and El Salvador, which gained more than eight NI points. Unfortunately, the nutrition situation in Haiti, the country with by far the lowest NI in the region (66.5 in 1997), remained virtually unchanged from 1981 to 1997.

Concerning **Eastern Europe and the countries of the former Soviet Union**, data availability permitted NI calculation for only a very limited number of states and points in time[7]. As Figure 5.2-2 shows, the nutrition situation in Armenia and Azerbaijan - both situated in the Caucasus region and recently involved in violent conflicts - and in the Kyrgyz Republic and Uzbekistan can be classified as mixed according to their NI scores, which range from 86 to 90. Two successor states of the former Yugoslavia (Croatia and the present Yugoslavia), Kazakhstan, the Russian Federation and

[6] Economic growth is even larger if the purchasing power of incomes is taken into account (World Bank 2000).

[7] NI values for 1981 and 1992 could not be determined for some countries, because they were not in existence in their current borders before the beginning of the 1990s.

Romania in 1992 reached comparatively better NI values above 94 that indicate a satisfactory, though not good, nutrition situation.

Summing up, it may be said that the North African and Near Eastern region, just as in Latin America and the Caribbean, experienced small to medium NI improvements from 1981 to 1997 and achieved a fair or mixed nutrition situation – with the exception of Haiti in the Caribbean and Yemen in the Near East. The few states of Eastern Europe and the former Soviet Union for which NI calculations were possible also showed good or mixed NI scores. The nutrition situation is still bad in South Asia and no better than middling in most of South-East Asia but, apart from the depressing situation in Afghanistan, there are very promising upward trends in this region. In Sub-Saharan Africa, the nutrition situation is bad or even extremely bad in about three quarters of the countries (the NI being below 80.4), though some West African states experienced large NI increases over the past two decades. Moreover, in large parts of Eastern and Southern Africa and in the West African state of Liberia, the quite bad nutrition situation in the beginning of the 1980s has deteriorated even since then. Therefore, not only in terms of absolute NI levels, but also with respect to recent trends, Sub-Saharan African nations give rise to the most serious concern about food security and nutrition.

6 NI and other International Indicators

This chapter deals with the relationship of the NI to other international indicators and thus seeks to answer research question three (see Chapter 1.2). As has been stated in Chapter 2.1.2, it would be of little use to calculate a Nutrition Index that is so highly correlated with existing indicators with broad data availability (referring to nutrition and food security only or to development in general) that the information contained does not provide any additional insights. Thus, the NI is tested for redundancy vis-à-vis its components and the main international development indicators, GNP per capita and HDI. In contrast, high correlations between the NI and indicators with scanty data availability are desirable, if the latter are meaningful for food security and nutrition. This pertains to indicators of micronutrient deficiencies, which could not be included in the NI due to lack of data, and to measures of poverty and distribution. The importance of analysing the sensitivity of the NI to poverty and inequality is rooted in the frequently voiced critique that indicators referring to national aggregates tend to conceal inequality and reveal little information on absolute poverty. Regarding indicators with a small database that should be largely reflected by the NI, redundancy is a favourable characteristic since it means that the NI is able to capture the included information without explicit consideration of these variables.

The calculation of rank correlations is the main methodological tool for analysing the relationship of the NI to other indicators in the present chapter. Rank correlations are chosen for this purpose because they entail a range of advantages in comparison to absolute correlations and to regression analysis:

- non-parametric measures (such as the Gini coefficient) can be easily incorporated into the analysis and the results are comparable to rank correlations for parametric measures
- in contrast to absolute correlations (expressed by Pearson's correlation coefficient), rank correlations can also be sensibly applied if the relationship of the indicators in question is non-linear (for example, of logarithmic form)[1]
- different from regression analysis, correlation analysis does not presuppose a uni-directional causal relationship between the variables under consideration which, e.g., does not exist between the NI and the HDI

---

[1] This does not hold, however, for a U-shaped functional relationship (Gujarati 1995).

- the relationship to indicators with scanty data availability (such as poverty measures and the Gini coefficient) can be conveniently investigated for subsamples of the total data set, whereas these variables can only be included in a multivariate regression at the price of considerably reducing the number of cases, and of diminishing degrees of freedom.

Furthermore, the Spearman rank correlation coefficient is the common measure of redundancy. The problem of specifying threshold levels of redundancy has already been touched upon in Chapter 2.1.2. Following roughly McGillivray and White 1993, a Spearman rank correlation coefficient of 0.90 and more is defined as 'Level 1' redundancy here. If the rank correlation coefficient is in between 0.70 and 0.90, this is referred to as 'Level 2' redundancy[2]. However, describing the statistical relationship of indicators by means of rank correlations may miss the explanatory value of 'outliers', i.e. countries that have considerably higher or lower NI values than would be expected according to the average relationship of the NI with other indicators.

Starting with a consideration of food security and nutrition indicators – including the NI components and micronutrient deficiencies – the analyses of the present chapter are continued with measures of poverty and distribution. The last section is dedicated to the relationship between the NI and development indicators. This part goes beyond rank correlation analysis by exploring the functional relationship of the NI to GNP per capita and HDI, respectively, and by seeking reasons for the outlier positions of selected countries.

6.1 Food Security and Nutrition Indicators

The list of food security and nutrition indicators that is examined with reference to the NI in the following is necessarily incomplete. The anthropometric indicators for wasting and stunting are not considered - they are not included in the Nutrition Index, but are believed to be roughly represented by the prevalence of underweight. Dietary energy supply per capita and the indicator for the depth of undernourishment (newly introduced with FAO 2000a) are likewise omitted, since dietary energy supply is very closely related to the percentage of undernourished (see the model in Chapter 4.1.1),

[2] The threshold value of 0.70 relies on the conviction that a new index is redundant if more than half of its variation is accounted for by an existing indicator. The formal interpretation of that claim is that the coefficient of variation is not less than 0.50, which corresponds to a correlation coefficient of 0.70 (approximately the square root of 0.50) (McGillivray and White 1993).

and the same is expected for the depth of undernourishment[3]. The analysis is restricted to the components of the NI, which evidently have the same or a broader data availability than the index itself, as well as to indicators of micronutrient deficiencies and the Hunger Index designed by Douglas Hicks (Bread for the World Institute 2001), for which only limited data exist. Calculating rank correlations for the NI and its components is used as a test of the composition of the index.

6.1.1 NI Components

The main aim of this subchapter is to test the redundancy of the NI vis-à-vis its partial indicators. But first, the relationship of NI components to each other is analysed, providing a test for the composition of the index (compare McGillivray and White 1993 about the HDI). According to Noorbakhsh (1998), one characteristic of a 'good composite index' is the fact that its components are not highly correlated – a statement that is in line with the requirements set forth by Szilágyi (2000) and Ryten (2000) (compare Chapter 2.3.1). For the pooled sample of NI 1992 and NI 1997, the Pearson correlation coefficients of the underlying indicators have already been calculated in the course of principal components analysis (see Chapter 4.3). In Table 6.1.1-1, Spearman rank correlation coefficients are shown for the full cross section sample - NI 1997 values were supplemented by NI 1992 values of the countries for which more recent NI scores could not be determined[4]. The data of the underlying indicators were matched with the NI data according to reference years. Subsamples were classified by NI level in the same manner as in Chapter 4.3. Due to its construction (compare Formula (4.1)) the NI rises with improvements in the nutrition situation, whereas its components tend to fall, which induces negative correlations for the NI with its partial indicators. Therefore, the threshold levels for redundancy (0.70 and 0.90) are applied for the absolute values of the coefficients, because an inversion of the ranking indicated by negative coefficients does not imply any information gain or loss. As the correlation coeffi-

---

[3] The depth of undernourishment is equivalent to the average dietary energy deficit of the undernourished (FAO 2000a). Since this number is derived from the same hypothesised distribution of dietary energy intakes and from the same dietary energy requirement cut-off as the percentage of undernourished, a close relation with this indicator can be assumed (similar to the association between the incidence of poverty and the poverty gap for a given poverty line).

[4] For the following countries, NI 1992 was used because lack of data prevented the calculation of NI 1997: Burundi, Congo (Rep.), Cuba, Djibouti, Ecuador, Ethiopia, Gabon, Guinea, Iraq, Liberia, Mexico, Mongolia, Morocco, Namibia, Panama, Paraguay, Romania, Saudi Arabia, Sierra Leone, Somalia, Swaziland, and Trinidad and Tobago.

cients show, Level 1 redundancy is not reached for the components of the NI vis-à-vis each other. The threshold level of 0.70 is only surpassed in the full sample for the relationship between under-five mortality and both undernourishment and underweight in children, meaning that the under-five mortality indicator is 'Level 2 redundant' as compared to undernourishment and underweight, and vice versa.

Table 6.1.1-1 Spearman rank correlation coefficients of underlying indicators for the full sample and subgroups classified according to NI level (p-values in brackets)

	Undernourished	Underweight
Full Sample (n=106)		
Underweight	0.647 ***	-
	(0.000)	
Under-five-mortality	0.719 ***	0.747 ***
	(0.000)	(0.000)
Low NI Group (n=35)		
Underweight	-0.162	-
	(0.353)	
Under-five-mortality	0.346 **	-0.018
	(0.041)	(0.917)
Medium NI Group (n=35)		
Underweight	-0.319 *	-
	(0.062)	
Under-five-mortality	-0.351 **	0.137
	(0.039)	(0.431)
High NI Group (n=36)		
Underweight	-0.130	-
	(0.449)	
Under-five-mortality	-0.106	0.431 ***
	(0.538)	(0.009)

Source author's own calculation based on data described in Chapter 4.1

All coefficients for the subsamples at different NI levels are far from the thresholds of redundancy, ranging from –0.35 to 0.43. Only four out of nine coefficients are significant at all. Interestingly, significant correlations are found for different indicator pairs across the subgroups. The contribution of the three indicators to the overall information content of the NI is therefore not uniform at different NI levels. For example, the underweight indicator contributes more additional information vis-à-vis the under-five mortality indicator at low and medium NI levels than in the high NI group, because a significant rank correlation of both indicators exists only in the latter country group (the correlation coefficient amounts to 0.43). Some of the correlations are even negative and significant, meaning that there is a tendency for the values of one indicator to be the higher, the lower the values of the other indicator. This is the case for the combination of undernourishment with under-five mortality and underweight in children with correlation coefficients of –0.35 and -0.32, respectively, for the medium NI subsample.

Table 6.1.1-2 Spearman rank correlation coefficients of the Nutrition Index and underlying indicators (full sample and subgroups)

Underlying Indicators and Nutrition Index	NI *(n=106)*	Low NI *(n=35)*	Medium NI *(n=35)*	High NI *(n=36)*
Undernourished				
Rank Correlation Coefficient	-0.914 ***	-0.759 ***	-0.348 **	-0.620 ***
P-Value	(0.000)	(0.000)	(0.040)	(0.000)
Underweight				
Rank Correlation Coefficient	-0.882 ***	-0.409 **	-0.665 ***	-0.613 ***
P-Value	(0.000)	(0.015)	(0.000)	(0.000)
Under-five mortality				
Rank Correlation Coefficient	-0.862 ***	-0.540 ***	-0.240	-0.318 *
P-Value	(0.000)	(0.001)	(0.165)	(0.059)
Underlying Indicators and Partial Nutrition Indices	NI *(n=106)*	Low NI *(n=35)*	Medium NI *(n=35)*	High NI *(n=36)*
Undernourished and $NI_{2,3}$				
Rank Correlation Coefficient	-0.714 ***	-0.023	0.413 **	0.138
P-Value	(0.000)	(0.898)	(0.014)	(0.421)
Underweight and $NI_{1,3}$				
Rank Correlation Coefficient	-0.726 ***	0.073	0.215	-0.030
P-Value	(0.000)	(0.676)	(0.215)	(0.864)
Under-five mortality and $NI_{1,2}$				
Rank Correlation Coefficient	-0.799 ***	-0.241	0.127	-0.123
P-Value	(0.000)	(0.164)	(0.466)	(0.474)

Source author's own calculation based on data described in Chapter 4.1

In a second step, rank correlation analysis is applied to the NI and its underlying indicators (see Table 6.1.1-2). Level 1 redundancy is in fact reached by the indicator of undernourishment for the full sample with a correlation coefficient of -0.91. But the picture is different for the NI subgroups: only Level 2 redundancy is attained for undernourishment and low NI values (the rank correlation coefficient is -0.76); the other coefficients for the medium and high NI group show non-redundancy[5]. For both underweight in children and under-five mortality, rank correlation coefficients indicate Level 2 redundancy vis-à-vis the NI for the overall sample. Again, the picture changes if the subgroups are considered, where only non-redundant coefficients in between –0.24 and –0.67 are found. Whereas the relationship between under-five mortality and the NI is strongest in the low NI group, underweight in children has its closest correlation to the NI in the two upper groups.

However, as McGillivray and White (1993) have pointed out with reference to the HDI, correlating a composite index with its components leads to a partial correlation of the index with itself, and it may be 'overly demanding' to require non-redundancy in this case. Therefore, the calculation of a restricted index, omitting one component and correlating it with the excluded component, is a better test of redundancy of the component indicators vis-à-vis the index. This is also a further check of the composition of the index: if a component is highly correlated with a partial index excluding the component, its inclusion provides little additional information. Thus, partial NI values have been computed and correlated with the excluded component[6]. The results are shown in the bottom part of Table 6.1.1-2. The largest decline in absolute values of coefficients for the total sample as compared to the upper part of the table is found for the undernourishment component – the coefficient is -0.71 instead of -0.91 if this indicator is correlated with the partial NI. Undernourishment contributes most to the overall information content of the NI for the entire sample; it should be kept in mind that it is the partial indicator with the largest standard deviation. The correlation of undernourishment with medium NI values is also notable (0.41), since it is positive and has the sole significant value among the coefficients for the subgroups. The positive relationship can be explained by the negative correlations of the undernourishment indicator with the other two NI components (compare Table 6.1.1-1). According to correlations of underweight in children and under-five mortality with partial Nutrition Indices, Level 2 redundancy is found for the whole sample, though the coefficients (-0.73 and –0.80, respectively) are again lower than for these indicators and the complete NI. The correlation of under-five mortality to the partial NI stands out. In this respect, not only the comparatively low standard deviation of this indicator should

[5] This is partly the outcome of reduced sample size, see the comment on Table 6.1.1-2 below.

[6] The sequence of the indicators is as follows: undernourishment (component 1), underweight in children (component 2) and under-five mortality (component 3). A partial NI excluding undernourishment as the first component is labelled $NI_{2,3}$, and accordingly for the other two partial indices.

be remembered, but also the fact that it has the highest rank correlations with the other two indicators for the total sample (see Table 6.1.1-1). Therefore, more of its information content is captured by the partial index composed of the remaining two indicators than for undernourishment and underweight in children. For the subgroups, the under-five mortality and underweight component also contribute substantially to the total NI, which is indicated by low and insignificant coefficients ranging from –0.24 to 0.22.

In conclusion, it may be stated that the NI adds most to the information contained in its components if subgroups of the full country sample are considered. Surely the relatively lower correlation coefficients for the subgroups partly arise from the lower sample size, because rank correlations tend to decrease with smaller numbers of cases. But the drop in correlation coefficients cannot be solely accounted for by reduced sample size, as a comparison with the rank correlations between differently calculated NI versions in Chapter 4.3 illustrates, where correlation coefficients remain very high even if the total sample is split up. Regarding the size and significance of the coefficients for the total sample and the subsamples, the findings in the preceding section are quite similar to results of studies about the HDI (McGillivray and White 1993).

6.1.2 Micronutrient Deficiencies

The connections between the most prevalent micronutrient deficiencies and the indicators the NI is composed of have been outlined in the conceptual part (see Chapters 3.1 and 3.2). Consequently, low NI scores can be expected to reflect a multitude of vitamin and mineral deficiencies, though indicators of micronutrient deficiencies were not explicitly included in the index due to insufficient data availability. In case indicators of vitamin and mineral deficiencies are found to be redundant vis-à-vis the NI and/or one of its components, the requirement of maximal correlation of selected with relevant non-selected variables would be fulfilled (compare Chapter 2.3.1). Whereas the NI should not be redundant vis-à-vis its components - as has been emphasised in the previous Subchapter - the case is different for indicators of micronutrient deficiencies: here, redundancy would be a desirable characteristic of the NI. Vitamin and mineral deficiencies that are non-redundant vis-à-vis the NI and the partial indicators are suitable candidates for inclusion in the index if their data availability is enhanced. However, the assessment of the statistical relationship of indicators of micronutrient deficiencies to the NI and its components is of a preliminary nature, because it is hampered by the same circumstance that precluded the integration of these indicators: lack of data. The samples under consideration are much smaller than the full set of countries with NI values for the 1990s. Therefore, relatively low rank correlation coefficients can be expected simply due to reduced sample size. Another problem is the wide time span from which the data for the prevalence of micronutrient deficiencies are drawn, which prevents the proper matching of indicators according to reference years. Thus, the criteria for redundancy should not be applied too strictly.

Rank correlation coefficients were calculated for NI 1997 supplemented by NI 1992 and iron, iodine and Vitamin A deficiencies as indicated by anaemia in pregnant women, the goitre rate in children aged 6-11 years, and sub-clinical as well as clinical Vitamin A deficiency[7] to test the relationship between micronutrient deficiencies and the NI. In addition, correlation coefficients for these deficiencies and some single indicators of the nutrition situation were computed, including the NI components. This allows for comparison of the values and significance levels of rank correlations for micronutrient deficiencies and the NI with those of the single food security and nutrition indicators, which may serve as a yardstick. Calculating correlations with NI components also makes evident how the partial indicators of the NI contribute to significant correlations of micronutrient deficiencies with the NI as a whole. Spearman rank correlation coefficients are shown in Table 6.1.2-1. The number of cases is given in the second column.

Considering the correlation coefficients, one should bear in mind that dietary energy supply per capita and the percentage of undernourished refer to the whole population, whereas the other indicators are based on a sub-group, mostly children of a certain age or, in the case of anaemia, pregnant women (whose nutritional well-being, however, has strong effects on children's health and nutritional status)[8]. The weak correlations of micronutrient deficiencies with dietary energy supply and the percentage of undernourished, respectively, may be partly explained by these differing reference groups. Despite the incomplete coherence of reference groups and the years of measurement (see the bottom part of Table 6.1.2-1), notably high - and also highly significant - rank correlation coefficients are observed, in particular for clinical Vitamin A deficiency.

---

[7] Sub-clinical deficiency was detected by measuring low serum retinol levels, clinical Vitamin A deficiency was identified by observing eye signs like Bitot's spots or blindness (The Micronutrient Initiative/ UNICEF/ Tulane University 1998).

[8] Furthermore, the coefficients relating to dietary energy supply and the NI are negative, because the figures tend to be the higher the better the nutrition situation, whereas indicators of micronutrient deficiencies and the NI components increase in numerical value with rising deprivation.

Table 6.1.2-1 Spearman rank correlation coefficients between micronutrient deficiencies, the NI and single indicators of the nutrition situation

Deficient Minerals/Vitamins and Indicators (in %)	Number of Cases	Spearman Rank Correlation Coefficients with				
		Single Indicators				Nutrition Index (NI)
		Dietary Energy Supply (DES)	Under-nourished in Population	Under-weight Children under 5	Under-five Mortality Rate	
Iodine						
Goitre in Children aged 6-11 Years	85	-0.169 (0.123)	0.189 * (0.083)	0.242 ** (0.026)	0.172 (0.115)	-0.236 ** (0.030)
Iron						
Anaemia in Pregnant Women	67	-0.288 ** (0.017)	0.342 *** (0.005)	0.414 *** (0.000)	0.330 *** (0.006)	-0.408 *** (0.001)
Vitamin A						
Subclinical Deficiency[1]	35	-0.110 (0.528)	0.153 (0.380)	0.597 *** (0.000)	0.556 *** (0.001)	-0.474 *** (0.004)
Clinical Deficiency[2]	29	-0.505 *** (0.005)	0.488 *** (0.007)	0.454 ** (0.013)	0.756 *** (0.000)	-0.712 *** (0.000)

Source author's own calculation based on the latest data from the following sources and years:
anaemia in pregnant women: 1985-95 (World Bank 1997)
goitre rate: 1985-97 (UNICEF 1999)
Vitamin A deficiency: latest data from 1982-1996 (MI/UNICEF/Tulane University 1998)
DES: mean 1995-97 supplemented by mean 1990-92 (FAO 1999a)
undernourished in population: 1995-97 supplemented by 1990-92 (FAO 1999b)
underweight children: latest data from 1993-1998 supplemented by latest data from 1987-92 (WHO1997/2000a)
under-five mortality: 1997 supplemented by 1992 (UNICEF 1999)

Notes 1 low serum retinol level
2 eye signs
* significant at the 10% level
** significant at the 5% level
*** significant at the 1% level

Regarding the goitre rate in children, the highest absolute value of the correlation coefficients is found for underweight in children (0.24), closely followed by the NI. The overall weak correlations between the total goitre rate and the other indicators are not surprising, given the fact that iodine intakes are strongly determined by the existence of successful salt iodisation programmes and by geographical conditions like the iodine content of soils and water - factors that are not necessarily reflected in the NI and the single indicators considered above. However, goitre as defined here captures those existing rather mild forms of iodine deficiency which are without immediate dangerous consequences for health. Severe iodine deficiency manifesting itself in cretinism has almost been eliminated in recent decades (UN ACC/SCN 2000). Despite the weak correlation of goitre prevalence with the NI in its present form, inclusion of this indicator may therefore be given no high priority, even if data availability is expanded.

Concerning anaemia, the highest absolute value of rank correlations is again observed for underweight in children (0.41), the coefficient for the NI being of almost equal size. The coefficients referring to the other single indicators are also significant, but lower. Considering the strong negative impact of anaemia on well-being and the importance of adequate iron supply for maternal and child health, the non-redundancy of anaemia vis-à-vis the NI and its components suggests that inclusion of this indicator in the NI would be desirable. However, this would require improved availability of recent data in order to avoid considerable restriction of the country data set and undue expansion of the reference time spans of the NI.

With respect to the two indicators of Vitamin A deficiency, significant rank correlations (–0.47 and –0.71) are found for the NI. However, sub-clinical Vitamin A deficiency is more closely correlated to the proportion of underweight children and to the under-five mortality rate[9]. The absolute value of the rank correlation coefficient for clinical Vitamin A deficiency and the under-five mortality rate exceeds the coefficient referring to clinical Vitamin A deficiency and the NI. But nevertheless, for the quite limited sample of 29 countries for which rank correlations with clinical Vitamin A deficiency could be computed, Level 2 redundancy is even reached vis-à-vis the NI. This is an important characteristic of the NI given the especially debilitating nature of this severe manifestation of Vitamin A deficiency (if clinical Vitamin A deficiency goes untreated, it leads to blindness and raises mortality risk from infectious diseases). If Level 2 redundancy of the prevalence of clinical Vitamin A deficiency vis-à-vis the NI holds or increases for a much larger sample, the inclusion of an indicator of clinical Vitamin A deficiency will have no high priority. Indeed, this deficiency seems to be quite well reflected by the NI by means of its under-five mortality component, taking into account the small sample size and the incoherence of reference years. The high correlation with the NI can largely be attributed to the close connection between the under-five mortality rate and clinical Vitamin A deficiency, which amounts to 0.76.

[9] Sub-clinical Vitamin A deficiency is an indicator which should be handled with care, though, because of the peculiarities of retinol homeostasis which limit the reliability of measurement. Infectious diseases like measles cause a sudden drop in blood levels of Vitamin A, which is only harmful, however, if the reserves in the liver are near to depletion. The characteristic eye lesions that are signs of clinical Vitamin A deficiency are much more specific for severe lack of Vitamin A. However, clinical Vitamin A deficiency cannot take account of moderate deficiencies that cause serious, but unspecific symptoms (Biesalski 1995).

This finding underscores the importance of the under-five mortality component of the NI. In fact, research in several countries showed that providing vitamin A supplements reduced young child mortality by 20 to 40% (Latham 1997), which illustrates the role of Vitamin A deficiency as a cause of child deaths.

In conclusion, compared to the single indicators, the NI performs quite well in terms of correlations with micronutrient deficiencies. The absolute values of coefficients referring to the NI all exceed the absolute values of the lowest coefficients found for its partial indicators and the vitamin and mineral deficiencies considered. Clinical Vitamin A deficiency is the only indicator considered that shows redundancy (Level 2 redundancy) vis-à-vis a selected variable. In the long run, it would be desirable to include an indicator of iron deficiency in the NI, if data availability improves. Despite low rank correlations, integrating an indicator of iodine deficiency into the NI seems to be a comparatively less pressing issue for the reasons stated above.

6.1.3 Hunger Index (HI)

The Hunger Index (HI) was developed by Douglas Hicks as a tool to assess the magnitude of countries' hunger problems, and was released by Bread for the World Institute (2001) shortly after the Nutrition Index was first published (Wiesmann et al. 2000). The disagreement about conceptual and data issues concerning the measurement of hunger is the background for the design of the HI. The basic idea behind the HI is quite similar to the rationale for building the Nutrition Index: it is meant to be a single index that combines data relating to different aspects of hunger. The HI seeks to take into account availability, access and utilisation of food as three elements of food security. Availability of food is thought to be represented by dietary energy supply per capita, which is adjusted for access by using the Gini coefficient for the inequality of consumption. This combined measure is preferred to the percentage of undernourished estimated by FAO, because the FAO figures are believed to give no due consideration to disparities in food supply (see Chapter 6.2.1 for evidence that is contrary to this belief). The measure of access-adjusted food availability is combined with the percentage of children that are not stunted, which is regarded as an indicator of proper food utilisation. The resulting formula is as follows (Bread for the World Institute 2001):

$$(6.1) \quad HI = \frac{\log[0.00321\ DES - 2.86](1 - CGINI) + (1 - CSTU)}{2}$$

when	HI:	Hunger Index
	DES:	dietary energy supply per capita (in kcal)
	CGINI:	Gini coefficient of consumption
	CSTUNT:	prevalence of stunting in children (as a share of total children)

Evidently, the formula for HI calculation is more complicated and the HI is less transparent than the NI, since it is much more difficult to predict changes in the HI from variations in the underlying indicators. The way the upper limit of dietary energy supply is set, as well as the use of the Gini coefficient of consumption, is questionable (however, such criticisms will not be the focus of attention in the present section).

Table 6.1.3-1 Spearman rank correlation coefficients for NI and Hunger Index (HI), with coefficients for NI components shown for comparison

	NI *(n=48)*	Lower NI[1] *(n=24)*	Higher NI[1] *(n=24)*
Hunger Index (HI)			
Rank Correlation Coefficient	0.911 ***	0.784 ***	0.812 ***
P-Value	(0.000)	(0.000)	(0.000)
Undernourished			
Rank Correlation Coefficient	-0.880 ***	-0.683 ***	-0.793 ***
P-Value	(0.000)	(0.000)	(0.000)
Underweight			
Rank Correlation Coefficient	-0.852 ***	-0.326	-0.825 ***
P-Value	(0.000)	(0.121)	(0.000)
Under-five mortality			
Rank Correlation Coefficient	-0.791 ***	-0.643 ***	-0.698 ***
P-Value	(0.000)	(0.000)	(0.000)

Source author's own calculations based on data from Bread for the World Institute 2001, and on data described in Chapter 4.1

Note 1 The countries in the joint data set with both NI and HI values have been ranked by the NI and have then been divided into two groups of equal size.

Regarding the calculation of the HI, restricted data availability for the Gini coefficient of consumption is most limiting – the HI could be only be calculated for 49 countries (the reference period of the underlying indicators is not stated), whereas the NI is available for 106 countries in the 1990s. The redundancy of the HI vis-à-vis the NI can be tested for a sample of 48 countries for which data for both indices are available.

Due to the smaller number of cases, Spearman rank correlation coefficients presented in Chapter 6.1.1 for the NI and its components are not directly comparable to the results obtained for the NI and the HI. Thus, coefficients for the correlations of the NI with its partial indicators have been computed for the limited sample of 48 countries for comparison, see Table 6.1.3-1.

Taking into account the sample size and the results in Section 6.1.1, the coefficients for the NI and its components are quite high in their absolute values, ranging from –0.68 to –0.88, except for the correlation of underweight with the NI in the lower NI group (-0.32)[10]. But the really interesting finding is that the rank correlation coefficients for the NI and the HI consistently exceed the coefficients for the components in the total sample and the subgroups, with the single exception of the correlation of underweight in children with higher NI values. Given the fact that correlating an index with one of its components implies a partial correlation of the index with itself (and claiming non-redundancy may be 'overly demanding' for this reason, as has been stated in Chapter 6.1.1) and that there are no common partial indicators of the NI and the HI, this is a bit surprising[11]. Despite different underlying concepts and components, Level 1 redundancy is reached for NI and HI for the total sample, and Level 2 redundancy for the two subsamples of equal size. The conclusion that the information content of NI and HI does not differ very much according to the results obtained for the joint data set, seems in place. The NI, however, is not only more transparent and easier to calculate; it has also an edge over the HI by virtue of its much broader data availability.

6. 2 Indicators of Poverty and Distribution

Indicators referring to national aggregates have been frequently criticised for concealing inequality and poverty within countries. This is particularly an undesirable characteristic of measures of central tendency like GNP and GDP per capita, which have no natural upper limits. Identifying and reaching the poorest via effective targeting -

---

[10] This may be explained by the fact that the HI covers about the same range of countries as the NI, with Ethiopia and Burundi at the bottom end and Croatia and Jordan at the top of the list, but with less than half the total number of countries for which NI values could be calculated. Thus, sorting out countries without HI values from the NI data set leaves gaps in the NI ranking. This implies a higher similarity between the NI and NI components rankings, since the differences between subsequent NI values become larger.

[11] The leading role of DES both in the HI and in estimating undernourishment as an NI component, as well as the use of an anthropometric indicator in both indices, may partly account for this.

which requires adequate monitoring instruments – has been a rallying cry not only in the food security and nutrition community. The NI is a measure for the absence of nutritional deprivation, which is not intended to be an alternative measure of poverty or inequality (though care has been taken in its design to avoid insensitivity to aspects of distribution). However, other things being equal, countries with high levels of economic inequality and poverty are expected to show wider disparities of nutritional well-being among different population groups than countries with a more equal income distribution and less poverty. It is hoped that the NI will account for such gaps by assigning relatively lower scores to countries with higher income inequality and poverty levels. For this reason the sensitivity of the NI to poverty and distribution is tested in the following subchapters, though an eventual ability of the NI to reflect inequality and poverty cannot replace its disaggregation by region, class and ethnic group.

6.2.1 Measures of Distribution

The following section deals with distributional aspects of the NI. The analysis applied roughly follows the methodology used by Lüchters and Menkhoff (2000b) to detect the 'equidistributional bias' of HDI measurement in comparison to GDP per capita. Empirical investigation in this field is hindered by scanty data on distribution. The World Bank has recently published information on distribution – Gini coefficients and the shares of the quintiles and the lowest and highest 10% of the population in income and consumption – for 112 countries (World Bank 2000). As the change in distributional indicators is usually quite sluggish, countries with data originating from the ten-year time-span up to the NI reference year 1997 were included in the analysis. Only Rwanda had to be excluded because the most recently available data on distribution are from 1983-85. For 28 developed countries with data on distribution, hunger and malnutrition were considered to be no major concern (see auxiliary criteria defined in Chapter 4.1), and no NI values were calculated. The data set was further reduced by 12 countries, for which lack of data prevented NI calculation, leading to a raw data set of 71 countries. Again, NI 1997 values were supplemented by NI 1992 values for the countries for which no more recent information was available.

Following Lüchters and Menkhoff (2000b), the difference in ranks assigned to countries with regard to NI and GNP per capita is taken as a measure for analysing the specific information content of the NI. As GNP per capita can fluctuate quickly from year to year, the data for this indicator were matched in the following manner with NI values: an average value was formed for each country with the mean GNP per capita (in constant 1995 PPP$[12]) of the three-year period to which the percentage of the undernourished referred, of the GNP per capita for the year to which the under-five mortality rate referred, and of the GNP per capita for the year the underweight data were collected. This was done to take into account the different points and periods in time when the underlying data for the NI 1997 and NI 1992 were taken. The difference between 'Real GNP per capita (PPP$) rank' and 'NI rank' is termed 'NI DIFF'. Positive NI DIFF values indicate that the relative nutrition situation of that country is superior to its relative income level.

If the NI captures any equidistributional element that is not found in GNP per capita, this should be evident from a systematic relationship between the two indicators Gini coefficient (termed GINI) and NI DIFF, which are correlated via a non-parametric Spearman rank correlation. The correlation coefficient is –0.29 and significant (two-tailed) with a p-value of less than 0.05 (the result is also shown in Table 6.2.1-2 on page 115, together with coefficients for other measures of distribution). The negative sign indicates that positive NI DIFF values coincide with a more equal distribution. The corresponding coefficient of correlation for the difference between GDP per capita and HDI rank and the Gini coefficient was only –0.21 according to Lüchters and Menkhoff (2000a). This might suggest that the NI is more sensitive to distribution than the HDI, but it should be stressed that the result cited above is not directly comparable to the correlation coefficient obtained for NI DIFF, because

- a different set of countries was used in the analysis carried out by Lüchters and Menkhoff - the raw data set of the present analysis is smaller because the HDI is available for a larger set of countries than the NI
- data were taken from a different period of time in the analysis performed by Lüchters and Menkhoff, i.e. from 1993 (HDI and GDP per capita) and 1983-93 (Gini coefficient)
- Lüchters and Menkhoff (2000a) refer to GDP per capita instead of GNP per capita which is used here.

[12] Purchasing Power Parity Dollars. Because GNP per capita data are only available in current PPP$ on the latest World Bank World Development Indicators CD-ROM (World Bank 2000), GNP per capita in constant 1995 PPP$ was calculated from these data by means of the ratio obtained by dividing GDP at market prices in constant 1995 US$ through GDP at market prices in current US$.

The analysis carried out for NI DIFF and GINI can be repeated for the same raw data set and time period with more recent HDI data to allow for better comparison of the 'equidistributional biases' of the NI and HDI. Since the social indicators largely constituting the HDI are subject to slower changes than GNP per capita, the NI 1997 and 1992 values were simply matched with HDI 1997 and 1992, respectively. HDI 1992 was recalculated according to the methodology of HDI 1997 from data published in UNDP 1994. The correlation coefficient obtained for HDI DIFF (which is defined here as the difference of GNP per capita rank and HDI rank) and GINI is, with a value of –0.17, only slightly lower than the correlation coefficient reported by Lüchters and Menkhoff (-0.21, p-value < 0.05), but it is insignificant (p-value > 0.10). The deviating result may be attributed to the three points stated above. Additionally, the methodology of HDI calculation was changed in 1999, when the logarithmic transformation of GDP per capita was introduced with HDI 1997 (UNDP 1999)[13]. This might have changed the relationship between rank differences and the Gini coefficient as it was found by Lüchters and Menkhoff for the earlier version of the HDI from 1993. This assumption may be checked by means of using HDI* instead of HDI, which is calculated in the same way as the HDI, but without the GDP component. Thus, HDI* remained unaffected by the methodological innovation in 1999 which concerns only the GDP part of the index. In fact, the Spearman correlation coefficient for HDI* DIFF (which is computed analogously to HDI DIFF) and GINI is significant for the raw data set assembled for the NI, and has the same value of –0.26 that Lüchters and Menkhoff found in their analysis of HDI* and GINI for a different country data set[14]. It should be noted that even compared to HDI*, which has a stronger 'equidistributional bias' than HDI, the NI reacts more sensitively to income inequality according to the correlation coefficients obtained for GINI and HDI* DIFF and NI DIFF, respectively, though the difference in correlation coefficients may be insignificant (compare Table 6.2.1-2 on page 115).

---

[13] With this methodological shift, shortcomings of GDP transformation in earlier HDI versions (compare criticisms made in Lüchters and Menkhoff 1996 and Lüchters and Menkhoff 2000b) were eliminated, whilst the discounting of income was actually reduced at high income levels (see UNDP 1999, p. 159, and comments made by Szilágyi 2000).

[14] This result suggests that the newly introduced logarithmic transformation of GDP per capita has actually weakened the distributional sensitivity of HDI, and that interannual variation of the systematic relationship between HDI and the Gini coefficient, the difference in country selection, and the use of GNP instead of GDP in the present analysis are not mainly responsible for the lower Spearman correlation coefficient found for HDI DIFF and GINI.

As far as the relationship between NI DIFF and GINI is concerned, it may be of interest to see which component of the NI is responsible for its significant overall 'equidistributional bias'. It may be suspected that all three indicators summarised in the NI contribute to its sensitivity to distribution. The percentage of undernourished, especially, might contribute substantially in this respect, because it is estimated under consideration of a distributional measure (namely the assumed interhousehold variation of dietary energy intakes). To shed light on this issue, the NI is split up into the percentage of undernourished (PUN) and the partial index $NI_{2,3}$[15]. PUN DIFF is the difference of the 'percentage of undernourished rank' and the 'GNP per capita rank', and $NI_{2,3}$ DIFF is computed analogously to NI DIFF. The Spearman rank correlation coefficient for $NI_{2,3}$ DIFF and GINI is negative (-0.07) but not significantly different from zero (p-value = 0.54). In contrast, the correlation coefficient between PUN DIFF and GINI amounts to –0.40 (p-value = 0.001) and is even higher than the correlation coefficient observed for NI DIFF and GINI. This raises the question whether this strong correlation has been introduced into the PUN indicator during its estimation process. FAO reports that for countries without direct information on the interhousehold variation of dietary energy intakes (which constitute the majority of countries with data on undernourishment), this variation was estimated by means of income distribution data that are at the same time the source of Gini coefficient calculation (FAO 1996b). In a further step, the main pillar of the calculation of the percentage of undernourished – dietary energy supply per capita (DES) – is examined for its potential 'equidistributional bias'. The Spearman rank correlation coefficient between DES DIFF and GINI is –0.38 (p-value = 0.001) and only slightly below the coefficient obtained for PUN DIFF and GINI. The equidistributional element of the NI can therefore be mainly traced back to the notable correlation between DES as the main base of the PUN indicator, and the Gini coefficient. This statement is important with respect to the validity of the PUN indicator, which has been blamed for undue consideration of inequality (Svedberg 1998). These criticisms can be at least partly rejected by the results obtained, which reveal that distributional aspects are already largely reflected by the dietary energy supply per capita indicator[16].

---

[15] This partial index is only composed of the child malnutrition and under-five mortality component of the NI. It has already been calculated for the analyses in Chapter 6.1.1.

[16] In contrast to income, there are both lower and upper physiological boundaries for people's dietary energy intake. Overconsumption of dietary energy by the rich can therefore not fully compensate for

Figure 6.2.1-1 Distribution of Gini coefficients for countries with extreme NI DIFF

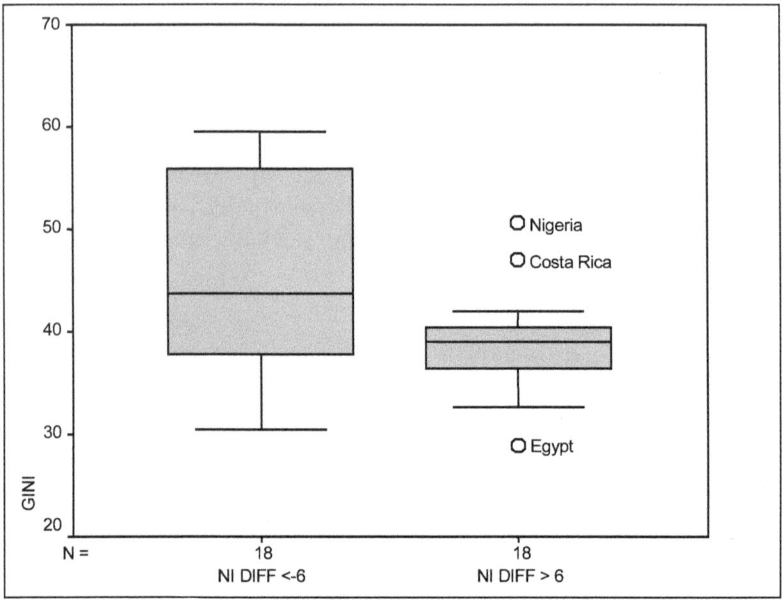

Source author's own presentation

Note The box covers the middle 50% percent of the data values between the upper and lower quartiles, the central line is the median. 'Whiskers' extend to the highest and lowest values, unless these are outliers (at a distance of more than 1.5 times the interquartile range from the margin of the box) or extreme values (at a distance of more than 3 times the interquartile range from the margin of the box). Outliers are represented separately by the 'O'-symbol.

The relation of NI DIFF and GINI is further examined by means of a box-and-whisker plot, in order to see if the significant correlation is eventually caused by a few outliers. First the 71 countries under review are arranged by NI DIFF and are divided into quartile groups. Figure 6.2.1-1 shows the groups at the extremes of the scale, each consisting of 18 countries. The whole distribution of Gini coefficients between the groups should be different if the correlation found for NI DIFF and GINI is meaningful. The left group is characterised by negative NI DIFF values between –21 and –7, the right group by positive values between 7 and 23. The median values of the Gini coefficients are 43.8 and 39.0, respectively, and the upper and lower boundaries of the boxes,

the shortfalls of dietary energy intake among the poor with respect to average values. This is an important difference to income, for which no natural upper limit exists.

which represent the interquartile ranges, have a different location, also. This allows for the conclusion that the significant correlation of NI DIFF and GINI is not due to a few outliers. In Figure 6.2.1-1, it leaps to the eye that the interquartile range of the left group is much broader than that of the right group[17].This suggests that a more equal income distribution is no guarantee for a relative nutrition situation that is superior to the relative income level. However, a high degree of income inequality seems to be a serious obstacle to the nutrition situation being favourable compared to average incomes. A closer look at individual countries may substantiate this view: the countries which have low Gini coefficients indicating little inequality, but NI DIFF values of −7 or below, are listed in Table 6.2.1-1.

Table 6.2.1–1 Countries with little inequality, but relatively low NI values as compared to GNP per capita in PPP$ (NI DIFF < -7)

Country	Gini coefficient		GNP per capita (PPP$)		Nutrition Index (NI)		GNP rank - NI rank (NI DIFF)
	value	rank	value	rank	value	rank	
Laos	30.4	4	1595	47	72.0	59	-12
Pakistan	31.2	5	1672	45	76.8	52	-7
Bangladesh	33.6	10	1352	53	65.6	68	-15
Sri Lanka	34.4	11	2710	35	79.0	50	-15
India	37.8	20	1969	42	74.4	54	-12

Source author's own calculations and World Bank (2000)

Note Ranks refer to a total of 71 countries. For ease of comparison, the Gini coefficient is ranked in ascending order, because lower values are thought to be more favourable, whereas GNP per capita and NI are ranked in descending order.

Laos, Pakistan, Bangladesh, Sri Lanka and India have much more favourable Gini ranks than GNP per capita and NI ranks. Whilst there is no straightforward explanation for the fact that the South-East Asian country Laos has an NI DIFF value of −12 in spite of its low Gini coefficient, pointing to a more equal income distribution[18], the four other countries under consideration are all in the South Asian region. Four out of five South Asian countries in the data set (all except for Nepal) combine low Gini co-

[17] Nigeria, that has a positive NI DIFF value in spite of a rather unequal income distribution, is an outlier of the right group. However, the case of Nigeria should not be taken too seriously in this analysis because the country is known for its weak statistical system.

[18] Laos, however, is the only country in the group listed in Table 6.2.1-1 for which the Gini coefficient is based on expenditure and not on income data. The Gini coefficient is, therefore, likely to be underestimated.

efficients with a nutrition situation that is considerably worse than relative GNP per capita would suggest. A possible explanation may be the heavy discrimination against women throughout South Asia, which is a serious obstacle to preventing child malnutrition – indeed, the prevalence of underweight is extraordinarily high in this region (Smith and Haddad 2000). Moreover, Sri Lanka has been plagued by a long-lasting civil war, and food supply is low as compared to other indicators (see Chapter 6.3.2). Croatia, the country with the lowest Gini coefficient in the sample, has an NI DIFF value of –2 (not shown in Table 6.2.1-1, because NI DIFF > -7). It seems that the war has not raised child malnutrition and mortality rates in Croatia (or at least not substantially), because they are still more favourable than for all other countries in the data set, but that food supply is comparatively low (see FAO 1999b). These examples illustrate that a relatively equal income distribution is no sufficient condition for positive NI DIFF values, because other factors like discrimination against women and war may counteract positive outcomes (compare Chapter 7, where the determinants of the NI are investigated in detail).

In the calculations presented so far, only the Gini coefficient has been considered as an indicator of distribution. However, this measure does not contain all relevant information with respect to distribution. Other indicators of distribution that reflect different aspects might produce heterogeneous results when used instead of the Gini coefficient for the computations carried out above. To check the robustness of findings, additional information on distribution is applied. Besides the Gini coefficient, the World Bank provides statistics on shares of income or consumption per quintile, and per highest and lowest 10%. The ratio of the highest quintile share to the lowest quintile share can be calculated from these data and is termed 'R20'. This measure has been applied, among others, by Lüchters and Menkhoff (2000b). Higher values of R20 indicate a less equal distribution.

The Spearman rank correlation coefficients for various ranking differences and indicators of distribution, including the Gini coefficient, are shown in Table 6.2.1-2. The coefficients for GINI and R20 are very similar and all have the expected negative signs. Except for two negative, but insignificant, coefficients relating to the correlation of $NI_{2,3}$ DIFF with the shares of the lowest 10% and 20% of the population, the coefficients consistently show the expected signs. The poorer 80% of the population (bottom four quintiles) can command comparatively larger income or consumption shares if the DIFF values are positive, that is if the rank of the indicator in question (NI, PUN, DES, HDI, HDI*) is better than the GNP per capita rank. The correlation coefficients for PUN DIFF and DES DIFF (0.40 for both) reveal that they have the strongest relationships with the income shares of the poorest (the lowest quintile), which become gradually weaker for the following three quintiles. For the NI, HDI and HDI*, the opposite tendency is observable: the correlation coefficients are gradually rising when proceeding from the lowest quintile to the fourth quintile. In contrast to the results for quintiles one to four, the income share of the highest quintile is negatively correlated with the DIFF values. There are only minor differences between the results for the bottom 10% and the bottom quintile on the one hand and the top 10% and the top quintile on the other (e.g., the correlation coefficient between NI DIFF and the share of the top quintile is –0.30 and rises to –0.34 for NI DIFF and the share of the top 10%).

Table 6.2.1-2 Spearman rank correlation coefficients for distributional indicators and ranking differences to GNP per capita (p-values in brackets)

	NI DIFF	$NI_{2,3}$ DIFF	PUN DIFF	DES DIFF	HDI DIFF	HDI* DIFF
GINI	-0.287 **	-0.073	-0.397 ***	-0.383 ***	-0.172	-0.260 **
	(0.015)	(0.543)	(0.001)	(0.001)	(0.151)	(0.028)
R20	-0.235 **	-0.012	-0.395 ***	-0.383 ***	-0.144	-0.229 *
	(0.049)	(0.919)	(0.001)	(0.001)	(0.231)	(0.054)
Lowest 10%	0.191	-0.039	0.378 ***	0.364 ***	0.129	0.205 *
	(0.111)	(0.746)	(0.001)	(0.002)	(0.284)	(0.086)
Lowest Quintile	0.218 *	-0.016	0.404 ***	0.396 ***	0.129	0.213 *
	(0.068)	(0.896)	(0.000)	(0.001)	(0.284)	(0.075)
Second Quintile	0.278 **	0.062	0.396 ***	0.377 ***	0.169	0.260 **
	(0.019)	(0.605)	(0.001)	(0.001)	(0.158)	(0.029)
Third Quintile	0.323 ***	0.113	0.378 ***	0.355 ***	0.185	0.268 **
	(0.006)	(0.348)	(0.001)	(0.002)	(0.123)	(0.024)
Fourth Quintile	0.394 ***	0.273 **	0.247 **	0.213 *	0.245 **	0.285 **
	(0.001)	(0.021)	(0.038)	(0.074)	(0.040)	(0.016)
Highest Quintile	-0.302 **	-0.089	-0.386 ***	-0.370 ***	-0.192	-0.276 **
	(0.010)	(0.459)	(0.001)	(0.002)	(0.109)	(0.020)
Highest 10%	-0.338 ***	-0.116	-0.383 ***	-0.364 ***	-0.197 *	-0.275 **
	(0.004)	(0.337)	(0.001)	(0.002)	(0.100)	(0.020)
Number of cases: 71						

Source author's own calculations based on data described in Chapter 6.2.1

The above checks for robustness by means of additional information on distribution confirm that the results obtained for the Gini coefficient are reliable and in no way extreme. Given the tightness of the correlation, NI DIFF performs better than HDI DIFF and HDI* DIFF, but is surpassed in this respect by PUN DIFF and DES DIFF. The rank correlation between $NI_{2,3}$ DIFF and the measures of distribution is astonishingly weak. Though this part of the overall NI does not therefore contribute to its sensitivity to equidistribution, it is relevant for the relationship of the NI to poverty measures, which is considered next.

6.2.2 Income Poverty

Measures of distribution are of a relative nature. They do not give us information on absolute levels of standards of living. This may be illustrated by the example of Brazil and Guatemala: the Gini coefficient is roughly equal for both countries (0.60 and 0.59, respectively), but Brazil's GNP per capita is about twice Guatemala's GNP per capita (World Bank 2000). Consequently, absolute poverty is much more prevalent in Guatemala than in Brazil, with about eight times more people living on less than 1$ a day (39.8% vis-à-vis 5.1% according to the data released in World Bank 2000[19]). In general, the percentage of the population living in absolute poverty is the larger, the lower per capita income.

For the present analysis, poverty measures based on international poverty lines have been used, because they are an attempt to hold the real value of the poverty line constant between countries, thus making international comparisons possible. The widely used 1$ a day standard is measured in 1985 international prices and adjusted to local currency using purchasing power parities (PPP)[20]. Additionally, poverty measures based on a 2$ a day standard, which represent a higher cut-off, are made available by the World Bank and used in the following analysis. The measures 'population below 1$ /below 2$ a day' are the percentages of the population living on less than the equivalents to 1$ and 2$, respectively, in 1985 prices (adjusted for PPP). The poverty gap is the average shortfall below the poverty line (counting the non-poor as having zero shortfall), expressed as a percentage of the poverty line. Strict comparability of poverty indicators across countries is not possible, because the primary household surveys differ in method and type of data collected. Though these problems are being reduced as survey methods improve and become more standardised, they still limit comparability (for further information, see World Bank 2000).

[19] The fact that the income share of the poorest quintile is even lower in Guatemala than in Brazil may further contribute to the much higher prevalence of income poverty.

[20] However, since PPP rates were initially designed for comparing aggregates from national accounts, but not for making international poverty comparisons, there is no certainty that an international poverty line reflects the same degree of deprivation across countries (World Bank 2000).

For the period 1983 to 1999, the World Bank has published data referring to international poverty lines for 77 countries altogether. 15 of these countries cannot be retained in the data set because there are no NI values available – among them are nine countries in transition, where the nutrition situation is assumed to be precarious, but data availability does not permit NI calculation (Belarus, Bulgaria, Estonia, Latvia, Lithuania, Moldova, Slovak Republic, Turkmenistan, Ukraine), and six countries for which problems of hunger and malnutrition are supposed to be of no major concern (Czech Republic, Hungary, South Korea, Poland, Portugal, Slovenia). For five countries, the data referring to the poverty line at 1$ are not very precise: it is only indicated that less than 2% of the population live below 1$ a day, and that the poverty gap is lower than 0.5%. In the following, these countries were assigned the same shared ranking place as if their indicator values were exactly 2% and 0.5%, respectively – in between the countries with 1.49% and 2.35% of the population below 1$ a day for the first indicator, and in between 0.46% and 0.52% poverty gap at 1$ a day for the second indicator.

Perfect matching of NI reference years and years of poverty data collection is not possible. The incidence and depth of poverty usually do not change overnight. As the extent of poverty not only follows long-term trends in distribution, but also depends on macro-economic performance, in times of crisis shifts may be less sluggish than for distributional indicators. In contrast to the procedure for measures of distribution followed in Subchapter 6.2.1, a ten-year time-span preceding the NI reference year seems therefore too long for the consideration of poverty indicators. Though some degree of reciprocity is claimed for poverty and the nutrition situation in the long run (UN ACC/SCN 1997), there is good reason to assume that poverty has stronger immediate effects on food security and nutrition than vice versa. Thus, changes in the incidence and depth of poverty may translate into NI changes within a few years (taking into account lagged effects on child malnutrition and the three-year period the estimation of undernourishment is based on). A time period ranging from the four years preceding the NI reference year and including the two years following it is thus chosen for the consideration of poverty data. This choice is based on the assumption that quick changes in poverty indicators are rare and that the two years following the NI reference year are still approximately representative for the extent of poverty in the NI reference year itself. For countries with the latest NI referring to 1997, information on poverty collected in the years 1993-1999 was therefore taken into account[21], and poverty data from 1988-1995 for countries for which the most recent NI is from 1992. For some countries with NI 1997, but no data on poverty from 1993 or later, NI 1992 was used to match poverty data from earlier years (this holds for Bolivia, the Gambia, Guatemala, Tunisia, Uganda, Uruguay and Zimbabwe)[22]. The matching of poverty measures, NI, HDI and the other indicators was carried out in the same way as described in Section 6.2.1. Similarly, the GNP per capita data needed for the calculation of ranking differences were chosen for the reference years of indicators – for the poverty measures, this was the survey year, for the percentage of undernourished the average of the years 1990-92 or 1995-97, respectively (depending on whether the country's NI dated from 1992 or 1997), and so on for the other indicators under review. Spearman rank correlations with the four poverty measures were calculated to identify the indicators with the closest relation to monetary poverty.

The rank correlations shown in Table 6.2.2-1 confirm the significant association between GNP per capita and poverty for two kinds of poverty measures (population below an international poverty line, expressed in PPP$, and poverty gap) and two different poverty lines (one dollar and two dollars). Compared to GNP per capita, the HDI and HDI* seem to entail no significant additional information on poverty, except maybe for the poverty gap at 1$ a day measure[23]. The tendencies observed for the indicators relating to national food availability (PUN and DES) are similar, but the rank correlations with the population below 1$ a day are even weaker than the respective

[21] There is only one country with poverty data from 1999 (Indonesia) and four countries with data from 1998 in the set. As far as inclusion of the most recent poverty data is concerned, Indonesia is a borderline case: after twenty years of steady decline in the incidence of poverty, the poverty rate more than doubled from 1996 to 1998 according to the national poverty line in the course of the Asian crisis, rising from 11.3 % to 23.2% (Said 2000). As the onset of the Asian crisis took place in 1997, its consequences are likely to be not yet fully reflected in NI 1997 due to the NI reference period extending over several years and to lagged effects. The alternative would be to drop Indonesia from the analysis, because poverty data published by the World Bank in previous years are not comparable to the data released in 2000 (the reason being updated PPP estimates, see World Bank 2000) – an option that is not chosen here.

[22] Rwanda and Botswana were omitted despite existing NI values and poverty data, because the surveys on poverty were conducted in 1983-85 and 1985-86, respectively, which was appraised to be not close enough to the reference period of NI 1992. Likewise, Panama was excluded from the raw data set because poverty data were acquired in 1997, and the latest NI value refers to 1992.

[23] The rank correlation coefficients for the poverty gap at 1$ a day and the two Human Development Indices are about 0.04 higher than the correlation between this poverty gap measure and GNP per capita. The correlation coefficients with HDI, HDI* and the other three poverty measures are either lower

correlation with GNP per capita. For underweight in children (CUW), no noteworthy rise of correlation coefficients can be observed in comparison to the absolute values of correlations with GNP per capita. In contrast, the rank correlations of poverty measures with under-five mortality (CM) increase by up to 0.09. As far as the $NI_{2,3}$ is concerned, it may be noted that the rank correlation coefficients for this partial index and the poverty measures exceed the respective coefficients with GNP per capita by 0.06 up to 0.09 (not shown in Table 6.2.2-1). This gain in information on monetary poverty in comparison to GNP per capita is even outperformed by the NI, as the correlation coefficients for the NI and the poverty measures are consistently larger than for all other indicators considered, including the NI components[24]. In comparison to the correlation coefficients with GNP per capita, the rise in absolute values is between 0.10 and 0.15.

Table 6.2.2-1 Spearman rank correlation coefficients for poverty measures, NI, NI components and development indicators (p-values in brackets)

	NI	CM	CUW	PUN	DES	HDI	HDI*	GNP p.c.
Population below 1$ a day (%)	-0.757 *** (0.000)	0.703 *** (0.000)	0.634 *** (0.000)	0.633 *** (0.000)	-0.613 *** (0.000)	-0.641 *** (0.000)	-0.632 *** (0.000)	-0.623 *** (0.000)
Poverty gap at 1$ a day (%)	-0.701 *** (0.000)	0.638 *** (0.000)	0.559 *** (0.000)	0.608 *** (0.000)	-0.597 *** (0.000)	-0.591 *** (0.000)	-0.591 *** (0.000)	-0.551 *** (0.000)
Population below 2$ a day (%)	-0.809 *** (0.000)	0.747 *** (0.000)	0.753 *** (0.000)	0.617 *** (0.000)	-0.582 *** (0.000)	-0.699 *** (0.000)	-0.673 *** (0.000)	-0.705 *** (0.000)
Poverty gap at 2$ a day (%)	-0.787 *** (0.000)	0.724 *** (0.000)	0.691 *** (0.000)	0.639 *** (0.000)	-0.616 *** (0.000)	-0.667 *** (0.000)	-0.648 *** (0.000)	-0.659 *** (0.000)
Number of cases: 59								

Source author's own calculations

than those with GNP per capita, or they have only slightly larger values than the correlation coefficients with GNP per capita (up to 0.02 more).

[24] In contrast to the NI, the Hunger Index (HI) designed by Douglas Hicks and presented in Chapter 6.1.3 contains explicit information on distribution. It is calculated by means of the Gini coefficient of consumption, which results in reducing data availability of the HI considerably. The question may be raised if this disadvantage is possibly outweighed by extraordinarily high sensitivity of the HI to poverty. To answer this question, Spearman rank correlation coefficients were calculated for the set of 38 countries for which the NI, the HI and the four poverty measures are available. The coefficients for the NI and the population below 1$ a day, poverty gap at 1$ a day, population below 2$ a day and poverty gap at 2$ a day amount to –0.76, -0.74, -0.77 and –0.77, respectively, (all p-values = 0.000). The corresponding coefficients for the HI and the poverty indicators are likewise highly significant, but a bit lower with values of -0.72, -0.72, -0.70 and –0.72 only. This suggests the conclusion that concerning sensitivity to poverty, the NI is slightly superior to the HI despite including only implicit information on poverty and distribution.

The significance of these findings is underscored by the coefficients in Table 6.2.2-2, where Spearman rank correlations for the four poverty measures and the differences between GNP per capita rank and the ranks of the indicators under review are shown (the ranking differences are termed NI DIFF, $NI_{2,3}$ DIFF, PUN DIFF, etc., and $NI_{2,3}$ is considered instead of CM and CUW, since it proved to be more closely associated with the poverty indicators than its components, see above).

Table 6.2.2-2 Spearman rank correlation coefficients for poverty measures and ranking differences of NI and other development indicators (vis-à-vis GNP per capita ranking)

	NI DIFF	$NI_{2,3}$ DIFF	PUN DIFF	DES DIFF	HDI DIFF	HDI* DIFF
Rankings						
Population below 1$ a day (%)	-0.288 ** (0.027)	-0.111 (0.405)	-0.055 (0.678)	0.007 (0.958)	-0.012 (0.925)	-0.011 (0.932)
Poverty gap at 1$ a day (%)	-0.318 ** (0.014)	-0.104 (0.435)	-0.118 (0.375)	-0.063 (0.635)	-0.060 (0.654)	-0.060 (0.651)
Population below 2$ a day (%)	-0.212 (0.106)	-0.145 (0.272)	0.093 (0.483)	0.162 (0.220)	0.024 (0.858)	0.036 (0.788)
Poverty gap at 2$ a day (%)	-0.264 ** (0.043)	-0.134 (0.313)	-0.008 (0.950)	0.055 (0.678)	0.007 (0.956)	0.012 (0.928)
Ranking differences to GNP						
Population below 1$ a day (%)	0.608 *** (0.000)	0.394 *** (0.002)	0.485 *** (0.000)	0.432 *** (0.001)	0.199 (0.131)	0.251 * (0.055)
Poverty gap at 1$ a day (%)	0.563 *** (0.000)	0.347 *** (0.007)	0.492 *** (0.000)	0.458 *** (0.000)	0.219 * (0.096)	0.265 ** (0.042)
Population below 2$ a day (%)	0.561 *** (0.000)	0.478 *** (0.000)	0.329 ** (0.011)	0.275 ** (0.035)	0.180 (0.174)	0.210 (0.110)
Poverty gap at 2$ a day (%)	0.600 *** (0.000)	0.431 *** (0.001)	0.451 *** (0.000)	0.395 *** (0.002)	0.214 (0.104)	0.260 ** (0.047)
Number of cases: 59						

Source author's own calculations

Note The ranking differences were calculated such that significant negative coefficients in the upper part of the table indicate that the respective indicator has a significantly stronger correlation with the poverty measure than GNP per capita (this holds for PUN, too, the only variable under review that decreases in value if the situation improves). In the bottom part of the table, positive correlations with ranking differences of the poverty measures indicate that the achievements with respect to the NI, the $NI_{2,3}$, low levels of PUN, DES, the HDI* and HDI tend to be the better relative to GNP per capita levels, the lower poverty is relative to GNP per capita.

The sole significant correlation coefficients in the upper part of the table are found for NI DIFF and apply to the relationships with three of the four poverty measures. They indicate that the NI in fact incorporates significantly more information on these poverty measures than GNP per capita (which serves for comparison by computing the ranking differences). The rank correlation coefficients with the other indicators range from –0.15 to 0.16 and are far from significant (with p-values ≥0.22). The fact that correlating NI DIFF ranks with the ranks according to the poverty measures does not result in higher correlation coefficients may be attributed to the strong systematic relationship between GNP per capita and income poverty (compare Table 6.2.2-1). The association of the NI and the other indicators with GNP per capita is actually eliminated by calculating ranking differences, whereas for the poverty measures no DIFF values are computed and the rankings according to these indicators reflect to a large extent the GNP per capita ranking. The correlations depicted in the upper part of Table 6.2.2-2 are, therefore, confused by the fact that the close association with GNP per capita is still retained for the poverty measures, but not for the other indicators.

This shortcoming of the present analysis can be avoided by expressing the poverty measures analogously to the other indicators, i.e. in the form of ranking differences. The bottom part of Table 6.2.2-2 shows the resulting Spearman correlation coefficients; the NI is far ahead of the other indicators in terms of high and highly significant correlation coefficients for all four poverty measure ranking differences, whereas only three values relating to HDI* and one referring to HDI are significant (two of these coefficients are only significant at the 10% level). Thus, if a country ranks very high according to its NI in comparison to its GNP per capita ranking place, this is likely to be associated with low levels of income poverty relative to GNP per capita. Again, the NI outperforms even its components $NI_{2,3}$ and PUN with respect to the closeness of this relationship[25].

Corresponding to the graphical analysis of the relation of NI DIFF and GINI in Chapter 6.2.1, a box-and-whisker plot is used to examine the distribution of ranking differences for the population below 1$ a day and GNP per capita and the lower and upper quartile of NI DIFF values (NI DIFF < -5 and > 6, respectively). Figure 6.2.2-1 shows that apart from the outliers Nigeria and Costa Rica, the countries in the upper

NI DIFF quartile all have positive ranking differences of the poverty measure vis-à-vis GNP per capita, whereas the countries in the lower NI DIFF quartile cover a broader range of predominantly negative ranking differences. (Note that positive ranking differences for the poverty measures indicate low levels of poverty relative to GNP per capita.) The difference in the median values of the quartiles is considerable. The results for the other three poverty measures are similar. Their box-and-whisker plots are not shown.

Figure 6.2.2-1 Distribution of differences between poverty measure rank (population below 1$ a day) and GNP per capita rank for countries with extreme NI DIFF

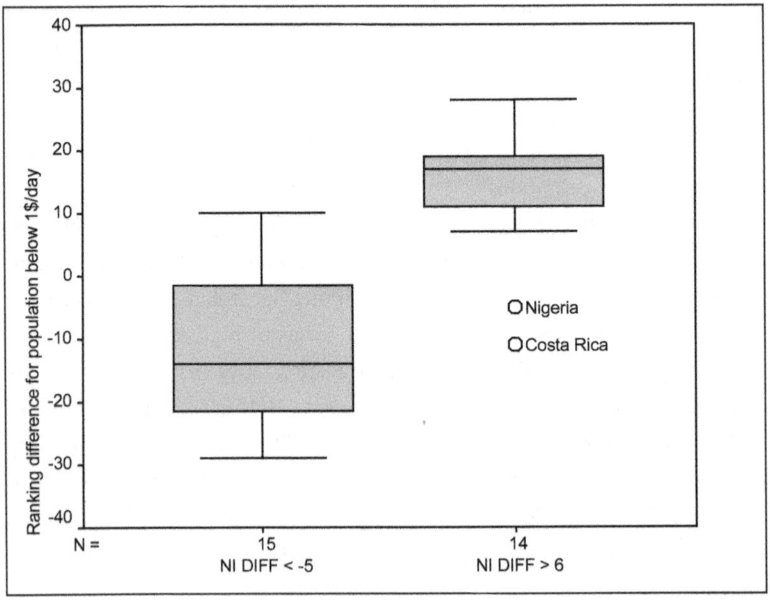

Source author's own presentation

Note The box covers the middle 50% of the data values between the upper and lower quartiles, the central line is the median. 'Whiskers' extend to the highest and lowest values, unless these are outliers (at a distance of more than 1.5 times the interquartile range from the margin of the box) or extreme values (at a distance of more than 3 times the interquartile range from the margin of the box). Outliers are represented separately by the 'O'-symbol.

---

[25] This suggests the conclusion that the factor isolated from the three NI components by means of principal components analysis in Chapter 4.3, that was used to derive the weights, is actually closely related to monetary poverty.

In the analyses presented so far, the NI clearly has an edge over the HDI and HDI* in terms of implicit information on monetary poverty measures, but it is not yet evident whether the NI encompasses *significantly* more information on poverty than the HDI and HDI*. To check for this, new ranking differences are calculated: the difference between NI rank and HDI rank (termed NI-HDI DIFF) and the difference between NI rank and HDI* rank (NI-HDI* DIFF) as well as the ranking differences of the poverty measures and HDI and HDI*, respectively.

Table 6.2.2-3 Spearman rank correlation coefficients for poverty measure ranking differences and NI ranking differences vis-à-vis HDI and HDI*, respectively (for NI 1997 supplemented with NI 1992)

	NI-HDI DIFF		NI-HDI* DIFF
Ranking differences to HDI ranking		*Ranking differences to HDI* ranking*	
Population below 1$ a day (%)	0.550 *** (0.000)	Population below 1$ a day (%)	0.501 *** (0.000)
Poverty gap at 1$ a day (%)	0.492 *** (0.000)	Poverty gap at 1$ a day (%)	0.436 *** (0.001)
Population below 2$ a day (%)	0.616 *** (0.000)	Population below 2$ a day (%)	0.596 *** (0.000)
Poverty gap at 2$ a day (%)	0.590 *** (0.000)	Poverty gap at 2$ a day (%)	0.563 *** (0.000)
Number of cases: 59			

Source author's own calculation

Computing the correlation coefficients for these ranking differences allows the comparison of the innovation of the NI vis-à-vis HDI and HDI* with respect to inherent information about income poverty. A positive correlation coefficient indicates that the NI rank being higher than the HDI rank corresponds to a relatively low level of poverty relative to the HDI. The values and significance levels of the rank correlations given in Table 6.2.2-3 in fact endorse that the NI incorporates significantly more information on income poverty than HDI and HDI* - a conclusion that again applies to all four measures of poverty used in the present analysis. The question may arise whether the use of NI 1992 values, which contain estimates of child malnutrition in some cases, may have biased the results of the correlation analysis in favour of the NI[26]. A closer look at the data reveals that an estimate of underweight in children has

[26] The model applied in UN ACC/SCN 1993 to estimate underweight prevalence actually encompasses 'per cent government social support' as an independent variable, meaning public spending on

entered the NI value of only one country in the present data set. A significant impact of the independent variables of the estimation model for child malnutrition in UN ACC/SCN 1993 on the ranking correlations is therefore virtually impossible. Nevertheless, the analysis is rerun with a limited data set that contains only data referring to countries with NI 1997 values. Using a partial data set for the analysis is at the same time a check for the robustness of the results; the number of cases is reduced from 59 to 42. Again, taking the relationship between differences in NI and HDI ranks on the one hand and HDI and poverty measure ranks on the other, the NI is found to include significantly more information on the various dimensions of income poverty, see Tables A 6.2.2-1 and A 6.2.2-2 in the Appendix.

Furthermore, it might be argued that the superiority of the NI vis-à-vis the HDI and HDI* in terms of a systematic relationship with poverty measures results from the longer reference period of the NI, which has an overlap with the years of poverty data collection. Though the social indicators included in the HDI, especially literacy and life expectancy, follow long-term trends, eventual quick changes in GDP per capita being also part of the HDI might weaken the connection of the HDI and the poverty measures with distant reference years. In a further analysis, the data set is reduced to 21 countries for which data on poverty were collected in 1991 or 1992 (matched with NI 1992 and HDI 1992) or in the years 1996 or 1997 (matched with NI 1997 and HDI 1997). Except for one poverty measure (poverty gap at 1$ a day) with a correlation coefficient of 0.37 and a p-value of 0.097, the correlations of NI DIFF values with ranking differences of poverty measures to GNP per capita retain significance at the 5% level despite the relatively small number of cases (see Table A 6.2.2-3). The comparative analysis of ranking differences of the NI and the poverty measures vis-à-vis HDI and HDI*, respectively, again reveals that the NI contains significantly more information on income poverty than HDI and HDI* (the results are shown in Table A 6.2.2-4 in the Appendix).

Summing up, the Nutrition Index is clearly more sensitive to the above considered measures of income poverty than GNP per capita, the HDI and HDI*, and even more sensitive than its components. In conclusion, of their findings relating to their investigation of distributional measures and the HDI, Lüchters and Menkhoff stated: 'The HDI thus represents a distribution-sensitive concept of welfare... . The introduc-

---

health, education and social security. Transfers from social security funds might contribute to the reduction of poverty.

tion and application of the HDI signals that the 1980s are gone, when "distributional and poverty issues were again pushed off the research and policy agenda" (Adelman and Robinson, 1989, p. 955)'. In analogy to these statements, it may be said that the NI entails not only a distribution-sensitive concept, but that it is also more sensitive to income poverty than currently prevailing development indicators. Taking into account the NI as an innovative measure in the field of food security and nutrition (instead of aiming solely at maximisation of GNP or HDI) would therefore urge policymakers to devote more of their attention and resources to poverty alleviation, and make it even more difficult to 'push poverty issues off the policy agenda'. A technical advantage of the NI is that it is available for 106 countries - 46 of which are without recent information on the proportion of people living below international poverty lines - and thus has a markedly wider coverage than explicit data on absolute poverty. Furthermore, income poverty measures neglect the availability of basic public goods and their role for standards of living, whereas non-monetary elements of poverty are considered in the NI via the impact of public services on child nutrition and survival. Taking the NI seriously would thus compel decision-makers to address poverty more effectively.

6.3 Development Indicators

The relation of the NI to commonly used development indicators with broad data availability is considered below for the entire data set of 106 countries. For redundancy analysis by means of calculating rank correlations, GNP per capita (in PPP$) and the HDI have been chosen as common and relevant indicators. Life expectancy, the component of the HDI which is meant to reflect nutrition together with health, is also considered. The data set with NI 1997 values supplemented by NI 1992 values (for countries without more recent information) is used, and the GNP per capita and HDI data are matched in the same manner with NI values as described in the preceding subchapter. As for the HDI, NI 1997 and 1992 were matched with life expectancy data of 1997 and 1992, respectively, because life expectancy is subject to much slower changes than GNP per capita. Eleven countries could not be included in the analysis regarding GNP per capita due to lack of data for the respective years[27] and for two countries no HDI values are available.

[27] This holds true for Afghanistan, Cuba, Djibouti, Iraq, Kuwait, Liberia, Libya, Myanmar, Rwanda, Somalia, Yugoslavia (Serbia and Montenegro).

Table 6.3-1 shows the Spearman rank correlation coefficients for the NI and the development indicators under review. The results are roughly comparable to those obtained in Chapter 6.1.1 for the NI and its components. Level 2 redundancy can be stated for the NI and all three of these indicators with coefficients ranging from 0.82 to 0.87, but Level 1 redundancy is not reached. The highest correlation coefficient in the subgroups amounts to no more than 0.53, and two of the coefficients are insignificant. Life expectancy is least correlated with the NI within the subgroups and is thus a poor correlate of the nutrition situation as measured by the NI. The question whether the NI provides additional insights as compared with the development indicators GNP per capita (in PPP$), HDI and life expectancy is clearly to be answered with yes if NI subgroups are considered.

Table 6.3-1 Spearman rank correlation coefficients for the Nutrition Index and development indicators (full sample and subgroups)

	NI	Low NI	Medium NI	High NI
GNP per capita (in PPP$)				
Rank Correlation Coefficient	0.872 ***	0.463 **	0.461 ***	0.201
P-Value	(0.000)	(0.010)	(0.007)	(0.270)
Number of Cases (n)	95	30	33	32
HDI				
Rank Correlation Coefficient	0.865 ***	0.527 ***	0.494 ***	0.283 *
P-Value	(0.000)	(0.001)	(0.003)	(0.099)
Number of Cases (n)	104	34	35	35
Life Expectancy				
Rank Correlation Coefficient	0.823 ***	0.180	0.382 **	0.302 *
P-Value	(0.000)	(0.301)	(0.024)	(0.073)
Number of Cases (n)	106	35	35	36

Source author's own calculation, partly based on data from World Bank 2000, UNDP 1994 and 1999

However, rank correlation analysis alone may not give full account of the informative value of the NI in comparison to development indicators like the HDI and GNP per capita. The question how well or how badly economic and human development are converted into nutritional well-being of countries' populations may be of particular interest. Therefore, the functional relationship of the NI to GNP per capita and the HDI is explored in the following section, and outlier cases are identified and explained.

6.3.1 GNP per capita

According to the conceptual framework in Chapter 3.1, the level of economic development is an important determinant of the nutrition situation. Therefore, GNP per capita may be entered as an explanatory variable in a univariate regression of the NI, which permits the identification of countries that do notably better or worse than would be predicted from their level of GNP per capita (possible endogeneity problems due to the feedback of bad health and nutritional status on learning and working capacity are not taken into account here). A linear regression with the logarithm of GNP per capita as an independent variable and the NI as a dependent variable yielded a good fit: adjusted R squared reached 0.74, which means that GNP per capita accounted for 74% of NI variation. Actual NI scores and predicted values are plotted against GNP per capita in Figure 6.3.1-1.

Figure 6.3.1-1 Actual and predicted values for the NI (1997 supplemented by NI 1992) plotted against GNP per capita

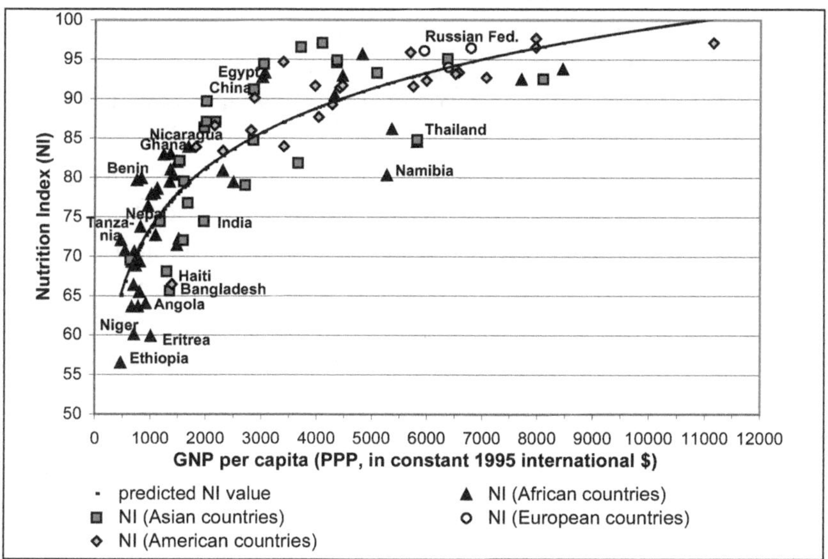

Source author's own presentation

Declining marginal returns to nutritional improvement of GNP per capita growth are evident. For example, a rise of GNP from 500 to 2,000 PPP$ per capita leads to an NI increase of about 15 points, whereas the same absolute rise from a level of 3,500 PPP$

to 5,000 PPP$ per capita results in a four point gain only. Large improvements in the nutrition situation can be brought about by rises in economic performance especially if the GNP per capita of a country is low. The disaggregation by regions shows that the bulk of African countries is concentrated in the low GNP per capita – low NI area of the graph. In contrast, the majority of Asian and Latin American countries can be found in the middle income class with higher NI scores. Of the three Eastern European countries for which data availability allowed inclusion in the analysis – Croatia, Romania and Russia – the latter two performed slightly better than their economic resources would lead one to expect.

The differences between actual NI scores and predicted values (the residuals) are listed in Figure 6.3.1-2 for the countries in developing regions. Of course, lacking willingness and ability of states to convert economic resources into nutritional improvements is not the only explanation for the divergences between actual and predicted NI scores. Residuals are surely to some extent the result of inevitable errors in the data and of random deviations. Too much emphasis should not, therefore, be put on small divergences from predicted values. The standard deviation of residuals is 5.46, and non-standardised residuals between –6 and 6 are shown in grey in Figure 6.3.1-2. The focus is on countries with larger negative or positive residuals, that are depicted in black and white, respectively, in the graphs.

On the African continent, quite a large number of countries have considerably worse nutrition situations than would be expected according to their GNP per capita – this applies in particular to the Southern African countries Namibia, Angola and Botswana, to Eritrea and Ethiopia in East Africa, and to Niger in West Africa. In contrast, the nutrition situation in Nigeria (which seems to have rather weak agricultural statistics), Benin, the North African states of Morocco and Egypt, and in Senegal and Tanzania is better than expected according to GNP per capita. In Asia, China and some countries in the Near East managed to convert economic resources quite well into nutritional well-being for their people. Substantially lower NI scores than predicted on the basis of GNP per capita could be observed for Thailand, Cambodia and Bangladesh. In Latin America and the Caribbean, only Jamaica with a quite good nutrition situation from the viewpoint of its economic development, and Haiti with a considerably lower NI than predicted, can be considered as outliers.

Figure 6.3.1-2 Residuals generated by the regression of the NI on GNP per capita

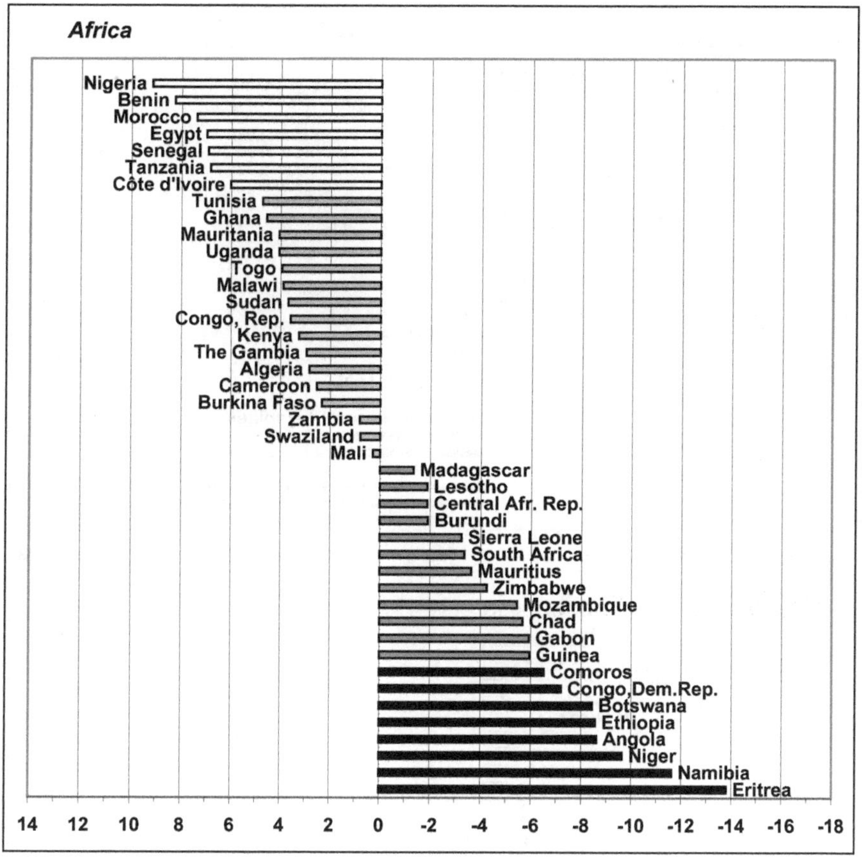

Source author's own presentation

Figure 6.3.1-2 (continued) Residuals generated by the regression of the NI on GNP per capita

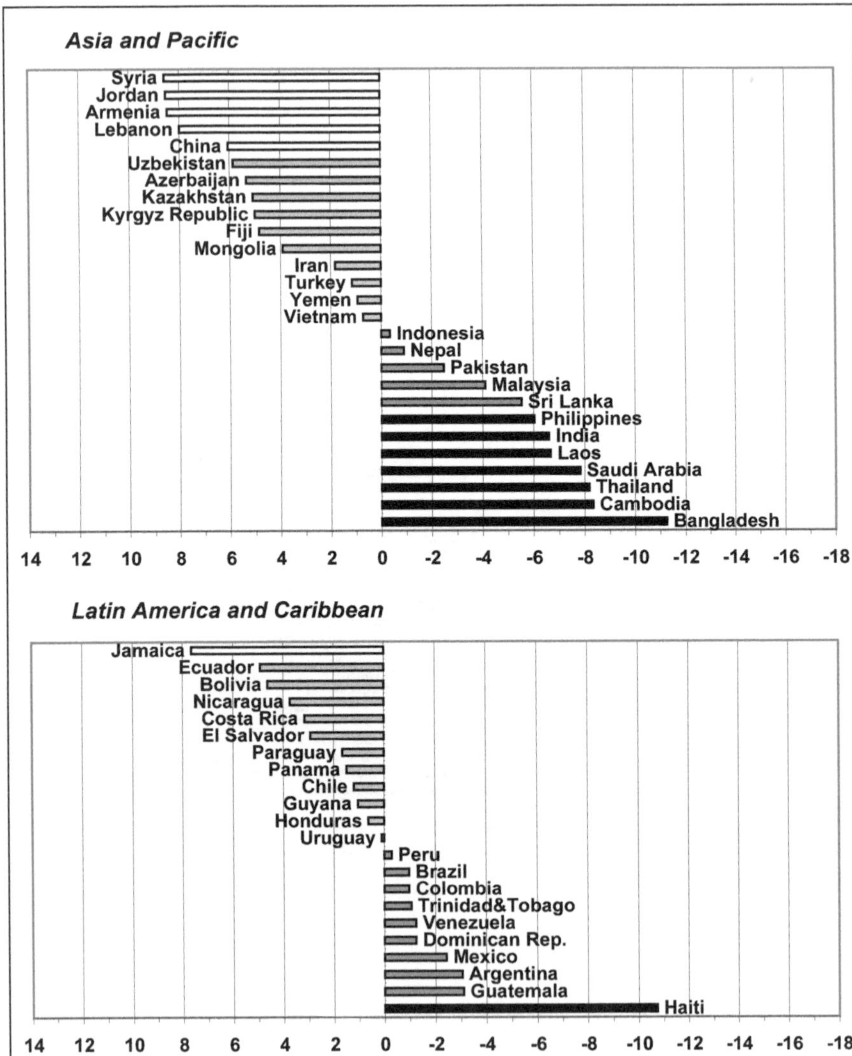

Source author's own presentation

Explaining the causes of particularly good or bad NI performance with respect to GNP per capita requires a closer look at the situation of individual countries. The following analysis of positive and negative outliers concentrates on selected low-income countries and a range of potential determinants of NI variation (see Table 6.3.1-1 on page 133). As starchy staples account for the largest share of dietary energy supply in poor countries (roughly 60-80%, see FAO 2000a) and low-income countries' capacity to purchase significant amounts of cereals on international markets is mostly limited, domestic cereal, roots and tubers production per capita is an important determinant of food security (compare Chapter 7.3.3). Furthermore, certain achievements in the fields of education (roughly represented by the literacy rate), access to and utilisation of effective health care (with immunisation against DPT as a proxy) and good sanitary conditions are necessary to reap the full benefits of food availability for child growth and health (Smith and Haddad 2000).

In both **Ethiopia** and **Tanzania**, undue policy regulations have restrained economic development for quite a long time (Webb and von Braun 1994, Schug 1991, Rwelamira and Kleyhans 1996). In war-ridden Ethiopia, fiscal and human resources have been diverted from the health and education sector to the military, and roads and medical facilities have been destroyed. In the virtual absence of irrigation technology, crop production and rural incomes remain vulnerable to low levels and high variability of rainfall which are found in several provinces. Frequent droughts, though not uniform across the country, contributed to crop failures and depletion of assets, which culminated in major famines, for instance in 1991, when more than six million people were affected (data taken from OFDA/ CRED 2000). The poor transportation infrastructure and policy restrictions on inter-regional trade impeded the flow of cereals and other staples between food-surplus and food-deficit regions (Webb and von Braun 1994). In addition, high rates of female illiteracy, poor water and sanitation conditions, and infant and child feeding problems are held responsible for the high prevalence of child malnutrition in Ethiopia (Pelletier et al. 1995). In Tanzania, agricultural production has responded positively to incentives created by economic reforms beginning in the mid-1980s (Rwelamira and Kleyhans 1996), and the country has been much less affected by drought than Ethiopia. Low rainfall has been no major problem until recently (major droughts with almost 10% of the population affected occurred in 1996 and 1998 according to OFDA/CRED 2000). In contrast to Ethiopia, Tanzania has only been indirectly affected by violent conflict since independence in 1961 (via refugees coming from other countries in the Great Lakes region), and has expanded the coverage of health programmes, primary education and sanitation to a remarkable extent, given the poverty of the country. This is indicated by literacy and immunisation rates, and reflected in child malnutrition and mortality rates that are quite favourable with respect to Tanzania's limited resource base (see Table 6.3.1-1).

The economic and political situation in **Angola** has been governed by almost uninterrupted warfare between competing parties since independence in 1975 (Feldbrügge 2000). As a consequence of population displacement, the disruption of markets and shortage of vital inputs, agricultural production and manufacturing industry nearly collapsed, whereas off-shore oil extraction remained largely unaffected by the fighting (Azam et al. 1994, IMF 1997). Revenues from the booming oil sector accrued to the government, and foreign exchange earnings were used to purchase grain on international markets (thereby compensating the shortfall in domestic cereal production to some extent, see data in FAO 2000b), but the lion's share of government resources has been spent for military purposes (up to 36.3% of government expenditure was allocated to 'defence and public order' in 1997 according to IMF 1999a). In the health and education sector, war has left a legacy of fragmented and low-quality basic services due to the destruction of schools and medical facilities as well as inadequate budgetary allocations (Rwelamira and Kleyhans 1996). In **Benin**, in contrast to Angola, recent agricultural development has been quite favourable: per capita production of starchy roots and cereals has risen considerably from 1970 to 1997 due to doubling of yields and an expansion of the area harvested. Whereas dietary energy supply per capita in Angola has fallen in the last three decades, it has increased in Benin from a similarly low level in 1970 by about 30% until 1997 (data from FAO 2000b). The better performance of Benin with respect to food supply is not quite as well reflected in the proportion of underweight children, probably because illiteracy and lack of sanitation are even more widespread in Benin than in Angola (and maybe because more malnourished Angolan children are likely to die), see Table 6.3.1-1. Child morbidity rates are indeed high in Benin, especially with respect to episodes of fever (Guédémé and Houndékon 1997).

Considering the South Asian countries included in the analysis, none of them is doing very well in terms of NI score relative to GNP level, the major reason being high malnutrition rates among children. Heavy discrimination against women throughout the region, which has a negative impact on female health, pregnancy outcomes and mothers' caring capacity, has been stated as one of the main causes of high malnutrition prevalence (Ramalingaswami et al. 1996, Smith and Haddad 2000).

Table 6.3.1-1 NI values of selected low-income countries and key indicators

Country	GNP per capita, PPP	NI 1997	NI components			Dietary energy supply per capita per day	Supply of staples per capita per year			Social indicators		
			Under-nourished	Under-weight children	Under-five mortality rate		Domestic production		Net imports of cereals	Adult literacy rate	Immunisation against DPT[4]	Access to sanitation[5]
							Roots & tubers[2]	Cereals[3]				
	(in intl.$)[1]		(in %)	(in %)	(in %)	(in kcal)	(in kg)	(in kg)	(in kg)	(in %)	(in %)	(in %)
Ethiopia	462	56	68	44	20	1674	15	108	16	33	13	10
Tanzania	468	72	40	31	14	2000	64	131	7	72	74	86
Difference	*6*	*16*	*-28*	*-13*	*-6*	*326*	*49*	*23*	*-9*	*39*	*61*	*76*
Angola	918	62	50	35	29	1900	59	36	41	45	41	40
Benin	838	80	15	29	17	2466	141	286	19	34	78	27
Difference	*-80*	*18*	*-35*	*-6*	*-12*	*566*	*82*	*250*	*-22*	*-11*	*37*	*-13*
Bangladesh	1352	66	37	56	11	2082	3	164	17	39	68	43
Nepal	1174	74	21	47	10	2315	10	232	2	38	78	16
Difference	*-178*	*8*	*-16*	*-9*	*-1*	*233*	*7*	*68*	*-15*	*-1*	*10*	*-27*
Haiti	1390	66	61	28	13	1837	25	51	55	46	35	25
Nicaragua	1818	84	31	12	6	2181	5	122	37	63	94	35
Difference	*428*	*18*	*-30*	*-16*	*-7*	*344*	*-20*	*71*	*-18*	*17*	*59*	*10*

Notes The numbers in the table have been rounded to whole numbers.

1 constant 1995 international dollar

2 dry equivalent; average dietary energy content of roots & tubers (dry eq., consumed as food) varies between 2400 kcal/kg in Tanzania and 5630 kcal/kg in Ethiopia (where enset, an indigenous tuber, plays an important role), calculated according to FAO 2000b

3 average dietary energy content of cereals (consumed as food): 3000-3400 kcal/kg, calculated according to FAO 2000b

4 proportion of one-year olds fully immunised against DPT (Diphteria, Pertussis, Tetanus)

5 proportion of population with access to adequate sanitation

6 data for Ethiopia refer to the period 1990-92, prior to the secession of Eritrea; the latest Nutrition Index is NI 1992

Source author's own presentation based on data from World Bank 2000, FAO 1999a, FAO 1999b, FAO 2000b, WHO 2000a, UNICEF 1999

The two poorest South Asian countries considered in the analysis are **Nepal** and **Bangladesh**, with the latter being slightly wealthier (see Table 6.3.1-1). Nevertheless, Nepal has performed better with respect to NI 1997 than Bangladesh. Whereas the value added by services accounts for 52% of GDP in 1996 in Bangladesh, the Nepalese economy is predominantly based on agriculture, with 42% of GDP being value added by agriculture in 1996 (the shares of industry and manufacturing in GDP are about the same in both countries) (World Bank 1998). Staple food production per capita is higher in Nepal, where the area harvested with cereals has been expanded by roughly 250% within the last three decades, though without significant increases in average yields (FAO 2000b). In Bangladesh, there has been no scope for an expansion of the area under cultivation due to the country's extremely high population density. Cereal yields have nearly doubled since 1970, but the growth of food supply could not outpace population growth (Ahmed et al. 2000), and dietary energy supply per capita has stagnated at low levels (FAO 2000b). Furthermore, the prevalence of very high levels of malnutrition in Bangladesh can be partly explained by adverse food habits: the conviction that women should eat less during pregnancy (Ministry of Health and Family Welfare, Government of the People's Republic of Bangladesh 1997) is especially likely to contribute to the worldwide uniquely high prevalence of low birth-weight of around 50% (UNICEF 1999), and infants with low birth-weight may never catch up the anthropometric shortfall at the very beginning of their lives (Ramalingaswami et al. 1996).

Concerning **Nicaragua** and **Haiti**, the comparison is more difficult because GNP per capita is considerably higher in Nicaragua (but these two countries are the single representatives of low-income countries from Latin America and the Caribbean in the sample). Haiti is the poorest country in the Western Hemisphere and has a long history of political crises and bad governance by an irresponsible elite that has enriched itself instead of serving Haiti's people (IMF 1999b, Gibbons 1999). Gross income inequalities are likely to persist until the present time, though no recent data on income distribution are available – it has been estimated that by the 1980s about '200 millionaires in Haiti enjoyed a life of unparalleled extravagance', whereas three quarters of the population existed in absolute poverty (Haggerty 1989). Deforestation and soil erosion, caused by high population density, extreme poverty and inadequate farming practices, have depressed agricultural production, together with inadequate pricing policies of the government: whilst cereal yields have stagnated, the yields of roots and tubers have even fallen slightly since 1970 (Icart and Trapp 1999, FAO 2000b). The quality and outreach of public services are poor: educational indicators (literacy and school enrolment rates) and access to sanitation and safe water are considerably lower in Haiti than in the average of low-income countries (classified according to World Bank criteria, IMF 1999b, World Bank 1999). The United States, the largest donor, has frequently interrupted aid flows because of alleged human rights violations, corruption and election fraud, and sanctions by the international community have worsened economic and social conditions during military rule 1991-1994 (Gibbons 1999). In contrast to Haiti, Nicaragua has a better NI value than its level of GNP per capita would lead one to expect (see Figures 6.3.1-1 and 6.3.1-2). However, the validity of this comparison may be questioned because GNP is assumed to be severely underestimated (by as much as 40%) due to the large informal economy and the weaknesses in the quality of the statistics. Internal political conflicts and civil war have dominated the country's history for quite a long time (EIU 1999), yet did not result in the severe deterioration of agricultural output per capita known from other war-torn countries. Despite the fighting, Nicaragua has had some of the best social indicators in Central America (with the exception of Costa Rica), but has fallen behind during the 1990s (EIU 1999). Yet the Nicaraguan health and education sectors are still in a better state than their Haitian counterparts, which is reflected in child malnutrition and mortality (of course, the impact of hurricane Mitch that hit Central America in late 1998 is not reflected in the NI 1997).

As the above analysis has shown, the NI contains relevant information on countries' nutrition situations that is not captured by GNP per capita. Besides considering the relationship of NI and GNP per capita, the comparative analysis of NI and HDI (including the identification of outlier cases) is another possibility for relating a country's nutrition situation to its level of development.

6.3.2 Human Development Index (HDI)

The Human Development Index (HDI) was designed in 1990 as a measure of development that has a broader focus than simply GNP per capita (UNDP 1990). It is composed of the logarithm of GNP per capita, life expectancy and the mean of literacy rate and school enrolment rates (UNDP 1999). All three rise with declining marginal rates when GNP per capita increases.

The connection of the Human Development Index (HDI) and the NI is illustrated in Figure 6.3.2-1 that depicts the close relationship of the two indices: the Pearson correlation coefficient amounts to 0.88 (being of similar size to the rank correlation coefficient presented at the beginning of this chapter). This high correlation for the overall sample is the result of the close connection of under-five mortality rate and life expectancy (indicators included in the NI and HDI, respectively) and of the strong impact of logarithmised GNP per capita and literacy rate - both part of the HDI – on the nutrition situation, as measured by the NI as a whole. Nevertheless, the two measures show a large divergence for some countries. Neither child malnutrition nor food insecurity are adequately captured by the HDI. Whereas the negative outliers at the bottom of the graph are nearly the same as in Figure 6.3.1-1, there are differences with respect to positive deviations from expected NI values: Tanzania's lead over Ethiopia in the health and education sector is well reflected in its much higher HDI and, with respect to HDI value, Tanzania is no positive outlier according to its NI. In contrast to this, Benin does only slightly better than Angola in terms of 'human development', which suggests that the difference in NI values can largely be attributed to disparities in food supply, (compare also Table 6.3.1-1).

Figure 6.3.2-1 Nutrition Index 1997 plotted against Human Development Index 1997

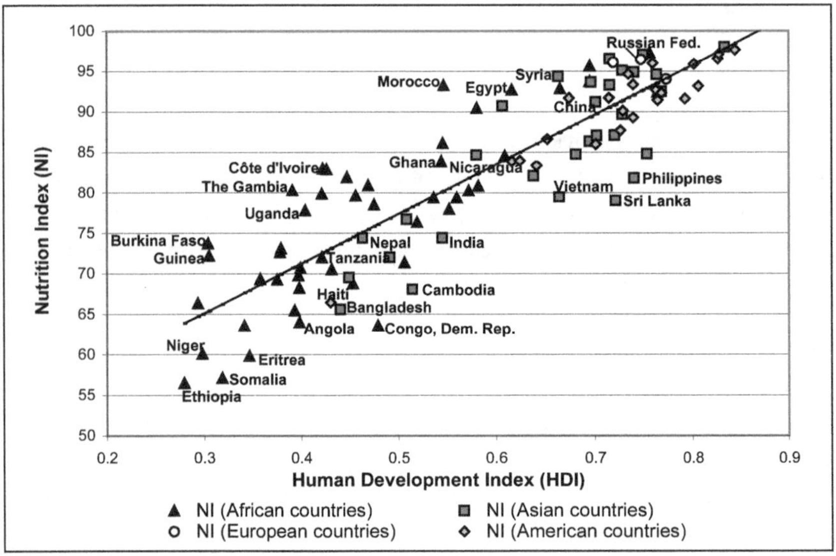

Source author's own calculation

Sri Lanka has a high HDI value as compared to its level of economic development, but its NI is considerably lower than would be expected according to its HDI. For the case of Sri Lanka, the measurement of human development has in fact a blind spot: though the country's achievements in the education sector and the high life expectancy are indeed notable and well reflected by the HDI, the HDI is unable to indicate the human misery caused by food insecurity and child malnutrition due to omission of respective indicators (compare the figures in Table A 5.1-1 in the Appendix and the discussion of human development in Ethiopia, Niger, Mozambique and Mauritania in Chapter 1.1). To quote Michael Hopkins: 'A second main conclusion of the report (the Human Development Report 1990, note of the author) is that respectable levels of human development are possible even at modest levels of income, although it cites an unfortunate example to help justify this – the war-racked island of Sri Lanka' (Hopkins 1991, p.1469). It seems that the NI gives a better assessment of the situation in Sri Lanka, though it is not intended to measure overall development. The inclusion of the percentage of undernourished in the NI can be held responsible for this feature – food supply is known to be adversely affected by conflict (Messer et al. 2001).

Another negative outlier in the comparison of NI and HDI is the Democratic Republic of Congo, that was plagued by a violent regime crisis from 1991 on, which turned into war in 1996 (according to the classification by HIIK 2000). The fighting made access to the main markets impossible, and occupying armies confiscated all food surplus available, thereby discouraging farmers from producing more than what they needed to survive. The fertile areas in the east that used to produce excess food sold to the western cities are now hardly able to feed the local population (Guiton 2001). The NI accounts for these detrimental consequences of the armed conflict on food security, whereas the HDI does not. In contrast to the Democratic Republic of Congo, Egypt is a positive outlier, see Figure 6.3.2-1. Part of the explanation can be found in the country's food subsidy system that has been a mainstay of long-term efforts to promote social equity and political stability. Despite problems like deficiencies in targeting and leakage, it is a major component of the social safety net for the poor, guaranteeing the availability of affordable staples, and has helped to reduce infant mortality and child malnutrition (Ahmed et al. 2001). Syria and Morocco are two other countries in the region of North Africa and the Near East that are positive outliers.

The impact of wars on the nutrition situation as measured by the NI, as well as possible reasons for the Near East and North Africa showing relatively good NI scores, are examined more closely in the subsequent chapter dealing with the determinants of NI variation.

In Boxes 6.3.1-1 and 6.3.1-2, the relation of subnationally disaggregated NI values to other indicators by province or federal state is described for China and India, including a comparison with HDI values for the Indian States.

Box 6.3.1-1 The NI for China's Provinces and other subnational indicators

In contrast to the weak relation stated for NI values and SDP per capita in India (compare Box 6.3.1-2), a significant correlation between net income and NI values is found for China (see also Figure B 6.3.1-1-1). The Pearson's correlation coefficient of logarithmised income and NI 1998 amounts to 0.68, implying that 46% of NI variation are explained by variation in income. Several explanations why the statistical relationships are different for China and India can be thought of. First, disposable income is a better proxy for people's command over resources than State Domestic Product per capita, and net income data are used for China for comparison with the NI. Second, incomes by province in China cover a wider range than SDP per capita in India – whereas income in Shanghai is more than six times the average income in Tibet, SDP per capita of the wealthiest Indian State is only about three times that of the poorest state. Third, the way the percentage of undernourished is aggregated from urban and rural estimates (with the urban population assigned a uniform value of 2%) may introduce a bias in favour of the wealthier provinces that are predominantly urban. However, the correlation between income and NI for Chinese provinces is not as strong as that found internationally for GNP per capita expressed in PPP$ and the NI (compare Chapter 6.3.1). Actually China, like India, is a very large and diverse country, and the income data are not adjusted by means of purchasing power parities.

Figure B 6.3.1-1-1 Nutrition Index 1998 plotted against net income for China's Provinces

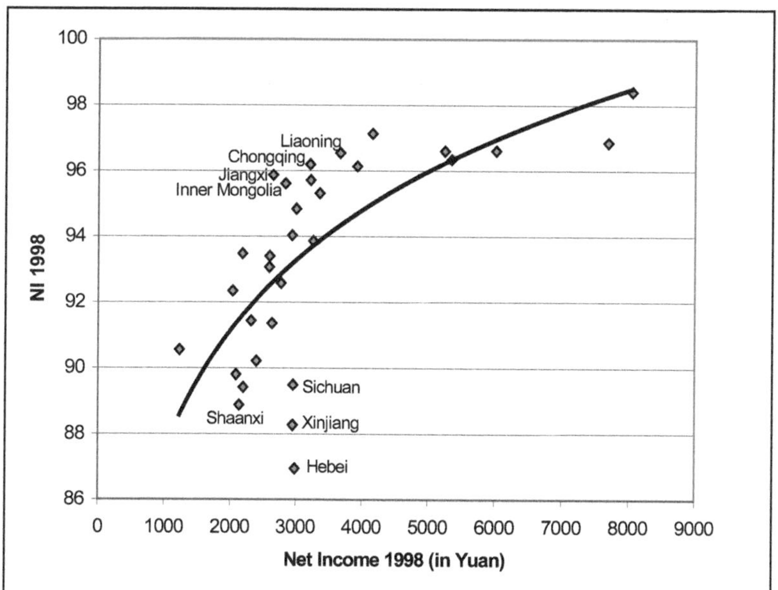

Source author's own presentation based on data described in Box 4.1.1-1

Box 6.3.1-1 (continued) The NI for China's Provinces and other subnational indicators

Regarding the outliers in Figure B 6.3.1-1-1, care should be taken in interpreting them due to the problems with data quality that are discussed in Box 4.1.3-1 in Chapter 4 - for example, the NI of Hebei is likely to be too low.

Table B 6.3.1-1-1 NI and selected indicators for China's Provinces

	Nutrition Index (NI) 1998	net income, urban and rural[1] (in Yuan) 1998	grain consumption per capita and year, rural[2] (in kg) 1998	illiteracy and semi-literacy[3] male (in %) 1997	illiteracy and semi-literacy[3] female (in %) 1997
1 Shanghai	98.41	8047	245	4.2	15.9
2 Jiangsu	97.13	4149	173	10.4	27.8
3 Beijing	96.86	7688	239	3.4	11.9
4 Tianjin	96.61	6009	267	4.7	14.9
5 Zhejiang	96.60	5243	255	10.8	25.9
6 Liaoning	96.55	3661	258	4.3	12.1
7 Guangdong	96.34	5341	241	3.3	15.8
8 Chongqing	96.20	3207	224	9.8	23.9
9 Fujian	96.14	3919	248	8.5	26.4
10 Jiangxi	95.87	2649	210	5.8	19.3
11 Jilin	95.71	3216	275	4.9	11.4
12 Inner Mongolia	95.60	2834	230	10.5	23.4
13 Heilongjiang	95.31	3355	276	5.5	13.0
14 Hubei	94.84	3001	301	8.3	22.0
15 Hunan	94.03	2937	268	5.8	17.0
16 Shandong	93.86	3260	262	13.2	32.0
17 Qinghai	93.48	2182	301	31.5	55.8
18 Anhui	93.39	2599	249	11.9	28.7
19 Guangxi	93.07	2595	244	7.4	23.2
20 Hainan	92.58	2773	260	7.1	21.5
21 Gansu	92.35	2034	253	17.0	36.8
22 Henan	91.43	2317	229	8.4	21.4
23 Shanxi	91.35	2638	232	6.2	13.7
24 Tibet	90.55	1232 *	244	44.2	62.8
25 Ningxia	90.21	2398	221	16.5	35.5
26 Guizhou	89.80	2088	244	14.1	38.8
27 Sichuan	89.48	2960	217	10.8	25.2
28 Yunnan	89.41	2197	233	16.5	34.1
29 Shaanxi	88.88	2138	255	11.4	23.5
30 Xinjiang	88.27	2954	244	9.2	14.0
31 Hebei	86.94	2988	257	8.3	20.3
China total	93.13	3184	249	9.6	23.2

Source author's own calculation and data sources cited below

Notes 1 aggregated from income data published in National Bureau of Statistics of China 1999
2 National Bureau of Statistics of China 1999
3 National Bureau of Statistics of China 1998
* refers to the rural population only, which comprises 89% of the population

Box 6.3.1-1 (continued) The NI for China's Provinces and other subnational indicators

Additionally, data availability at the provincial level is sometimes too scanty for explaining NI differences, though Table B 6.3.1-1-1 shows some variables that are found to have a statistically significant impact on the NI. However, in the comparison of Chongqing and Sichuan, two provinces that originated from the division of the former Sichuan - it is not obvious from the data in Table B 6.3.1-1-1 why the NI is six points higher in Chongqing than in Sichuan. The much lower prevalence of child malnutrition in Chongqing is mainly responsible for the difference in NI values. The levels of income, literacy and dietary energy supply per capita (in rural areas), though, are similar for the two neighbouring provinces. Important differences may exist with reference to public service provision, as Chongqing is more densely populated and more highly urbanised than Sichuan. Together with the eastern part of Sichuan, Chongqing belongs to the so-called Red Basin, an area of high agricultural productivity and population density, and the city of Chongqing is a huge industrial centre, one of China's booming cities. The western part of Sichuan is scarcely populated and has a cooler climate. Unfortunately, no data are available with respect to access to health services, sanitation and safe water. A more detailed investigation of NI differences is therefore left to further research.

Box 6.3.1-2 The NI for Indian States and other subnational indicators

The connection between the disaggregated NI for Indian states and per capita State Domestic Product (SDP) and the HDI is rather weak: rank correlation coefficients are 0.25 and 0.28, respectively, and are not significant. These results differ greatly from correlation coefficients obtained with international data. Besides the considerably larger variance of international data, limited comparability of SDP across states, which is rooted in different compilation methods, may be one reason. Furthermore, SDP per capita is expressed in Rupees and not adjusted for differences in purchasing power, although a certain variation in prices and purchasing power can be taken for granted in a country as large and diverse as India. This also affects the disaggregated HDI, one third of which is composed of transformed SDP per capita. In Figure B 6.3.1-2-1 and Figure B 6.3.1-2-2, the 1994 NI is plotted against SDP per capita 1993-94 and HDI 1995, respectively. The lines in the graphs depict the tendencies of the relationship. The positive and negative outliers are nearly identical for the comparison of the NI with both indicators. In Figure B 6.3.1-2-2 referring to the HDI, Kerala is less distant from the trend line, because its social indicators are very favourable in relation to its level of economic resources, which leads to a much higher HDI than the State Domestic Product suggests.

Figure B 6.3.1-2-1 NI 1994 plotted against SDP per capita 1993-94 for Indian States

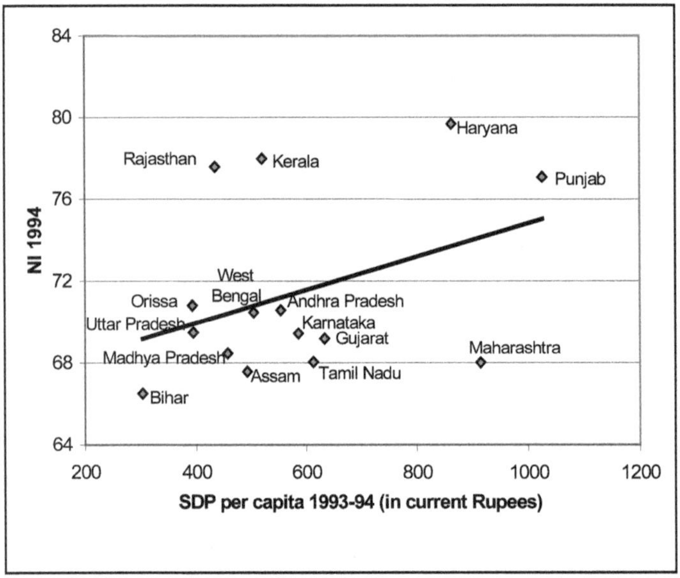

Source author's own presentation based on data shown in Table B 6.3.1-2-1

Box 6.3.1-2 (cont.) The NI for Indian States and other subnational indicators

Figure B 6.3.1-2-2 NI 1994 plotted against HDI 1995 for Indian States

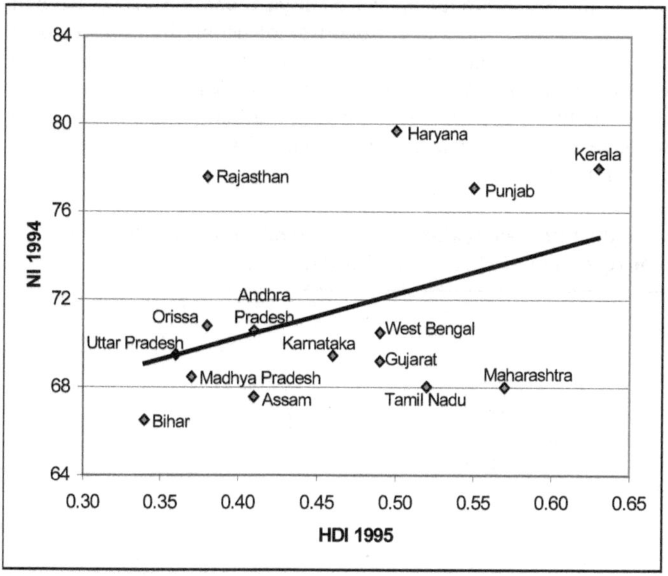

Source author's own presentation based on data shown in Table B 6.3.1-2-1

For the 1994 NI values of Indian states and the most widely used poverty measure - the percentage of the population living below the poverty line - the rank correlation coefficient is significant at the 5% level (p-value = 0.025) and amounts to 0.58. In so far as the national poverty line is adjusted to different local prices and commodity bundles, it is better suited to consider differences in purchasing power than SDP per capita. Furthermore, the proportion of people living below the poverty line is an indicator that is less prone to concealing inequalities than SDP or GDP per capita, because it is a cut-off-based indicator and no measure of central tendency. This means that it is not influenced by gains for the wealthiest in society. 1994 NI plotted against the proportion of poor people in Indian states is shown in Figure B 6.3.1-2-3 (see next page). Bihar, the poorest state, and Punjab, the wealthiest state according to SDP per capita, are not outliers with reference to the NI if their levels of poverty are considered. In Table B 6.3.1-2-1 on page 145, the ranking of Indian states according to 1994 NI is presented together with State Domestic Product, HDI, poverty and other data related to the nutrition situation[28].

[28] The inclusion of dietary energy supply per capita (DES) may seem unnecessary, because this indicator has already indirectly entered the NI via the estimation function for the percentage of undernourished. However, considering DES is helpful for assessing the relative contributions of food availability to the NI on the one hand, and factors such as adequate provision of public services on the other.

Box 6.3.1-2 (continued) The NI for Indian States and other subnational indicators

As already mentioned, in the wheat producing states Haryana and Punjab, which both have high State Domestic Products and a low proportion of people living in absolute poverty, the nutrition situation is relatively good compared to other Indian states. Neighbouring Rajasthan, too, has a minor incidence of poverty despite its low SDP per capita, and is ranked in the top four. The wheat growing northern states of Rajasthan, Haryana and Punjab have the highest dietary energy supplies in India, which is favourable for their overall nutrition situation.

Figure B 6.3.1-2-3 NI 1994 plotted against percentage of population below the poverty line in 1993-94 for Indian States

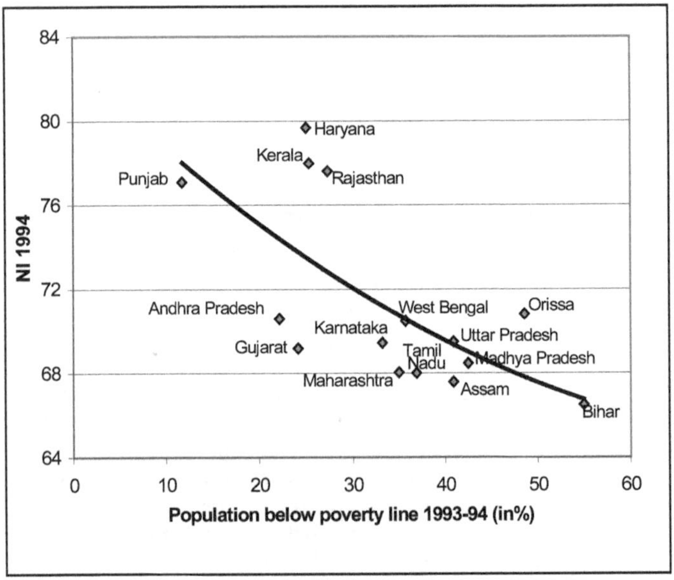

Source author's own presentation based on data shown in Table B 6.3.1-2-1

Though large parts of Uttar Pradesh also belong to the wheat surplus producing area, and it is ranked fourth according to dietary energy supply per capita, its NI is considerably lower. Reasons can be found in the large proportion of people living below the poverty line (ca. 41%) and in weaknesses in the educational sector (the literacy rate is well below the Indian average). These factors hamper the translation of high food availability into gains in child nutritional status and survival.

Kerala in South India is a somehow exceptional case among Indian states. Kerala has often been cited in the literature for its high achievements in the health and education sector. The good status of women as compared to the rest of India is also notable.

Box 6.3.1-2 (continued) The NI for Indian States and other subnational indicators

The uniquely high ratio of females to males is a crude reflection of women's enhanced survival chances in Kerala, with favourable consequences for women's caring capacity. Kerala is also doing very well in terms of the Nutrition Index. It should be noted, though, that Kerala's high NI value can mainly be traced back to low prevalence of child malnutrition and low under-five mortality. Regarding conversion of food availability into good nutritional status and health of children, Kerala has managed to do 'much with little'. Kerala could do even better if food supply could be enhanced, thereby reducing the quite high level of undernourishment.

Table B 6.3.1-2-1 NI, SDP per capita, HDI and selected indicators for Indian States

	Nutrition Index[1] 1994	State Domestic Product per capita[2] (in current Rupees) 1993-94	HDI[3] 1995	dietary energy intake per capita[1] (in kcal) 1993-94	population below poverty line[2] (in %) 1993-94	literacy rate[4] (in %) 1991	sex ratio[4] (females per 1000 males) 1991
1 Haryana	79.68	863	0.50	2517	25.1	55.9	865
2 Kerala	77.99	520	0.63	2057	25.4	89.8	1036
3 Rajasthan	77.60	435	0.38	2454	27.4	38.6	910
4 Punjab	77.08	1027	0.55	2429	11.8	58.5	882
5 Orissa	70.81	394	0.38	2310	48.6	49.1	971
6 Andhra Pradesh	70.58	554	0.41	2131	22.2	44.1	972
7 West Bengal	70.48	505	0.49	2291	35.7	57.7	917
8 Uttar Pradesh	69.48	395	0.36	2374	40.9	40.7	879
9 Karnataka	69.43	586	0.46	2154	33.2	56.0	960
10 Gujarat	69.18	633	0.49	2099	24.2	61.3	934
11 Madhya Pradesh	68.47	457	0.37	2245	42.5	44.7	931
12 Tamil Nadu	68.03	613	0.52	1985	35.0	62.7	974
13 Maharashtra	68.01	915	0.57	2050	36.9	64.9	934
14 Assam	67.57	493	0.41	2090	40.9	52.9	923
15 Bihar	66.50	304	0.34	2224	55.0	37.5	911
India total	70.40	588	0.45	2230	36.0	51.6	927

Source author's own calculations and data sources listed below

Notes 1 see Table B 4.1.3-2-1 in Chapter 4.1.3
2 Directorate of Economics and Statistics 1995
3 Srinivasan 2000
4 Census Data Online 2001

7 Determinants of NI Variation

A multivariate regression analysis treating the NI as an aggregate dependent variable can only be a crude attempt to identify the main overarching determinants of countries' nutrition situations. First, cross country studies are subject to general limitations which have been summarised by Smith and Haddad 2000. Concerns have been raised about data availability, reliability and comparability across countries and about subnational applicability of findings. Some variables relating to political decisions that are exogenous at the household level must be considered as endogenous at the national level (Smith and Haddad 2000). Second, modelling the NI as an aggregate has inherent limits. For example, dietary energy supply per capita (DES) has been shown to be an important determinant of child malnutrition in several cross country analyses (Smith and Haddad 2000, UN ACC/SCN 1993). However, employing DES as an explanatory variable in a regression with the NI as a dependent variable would not be acceptable, because it would duplicate the input into the equation of undernourishment, since DES has already entered the process of estimating undernourishment carried out by FAO (FAO 1996a). For deeper insights, modelling the components of the NI separately and considering their interactions is, therefore, inevitable – an approach that is, however, not pursued in the present study[1]. Third, a multivariate cross country analysis provides only a snapshot of economic, political, social and natural conditions and their outcomes in terms of food security and nutrition at a given point in time. The historical development and interplay of explanatory factors is not reflected in such models. Casting light on dreadful synergies between policy failures like economic mismanagement and violent conflict, natural resource degradation, and rapid population growth, is left to careful country level analyses, such as studies of the human misery in Ethiopia in the nineties (Schug 1991,Webb and von Braun 1994).

Nevertheless, merits of cross country studies are found in the use of variables that are only available at the national level, for example indicators relating to political systems or national incomes (Smith and Haddad 2000). Additionally, treating the NI as

---

[1] Estimating dietary energy supply per capita by country, the main base of undernourishment figures, necessitates the application of huge simulation models (see Alexandratos 1995, Rosegrant et al. 2001). Cross country analyses of child nutrition and survival have been carried out by Anand and Ravallion 1993, Subbarao and Raney 1995, and Pritchett and Summers 1996 for child mortality, by UN ACC/ SCN 1993, Rosegrant et al. 1995, Gillespie et al. 1996, Frongillo et. al 1997, Osmani 1997, and Smith and Haddad 2000 for child malnutrition, and by Klasen 1999 for child malnutrition and mortality, and will not be reproduced here. Thus the single indicators composing the NI are only taken into consideration for formulating hypotheses on explanatory factors, and for explaining the significance of variables in the aggregate NI model.

an aggregate dependent variable has not only limitations, but also advantages in comparison to a separate analysis of NI components. Modelling the NI as a whole can help to find overarching variables acting simultaneously and in the same direction on the single indicators summarised in the NI. Furthermore, as has been argued in Chapter 2.2, the impact of measurement error on the NI is believed to be weaker than the influence of measurement error on its components. This is likely to reduce the variance of the error term in the regression and to increase the explanatory power of the model (see Part 7.1.3 for further explanation).

In accordance with the conceptual framework laid out in Chapter 3, a set of hypotheses and related explanatory variables is described below. The operational measures of variables and data sources are detailed in Section 7.2, before testing them econometrically in Chapter 7.3.

7.1 Hypotheses on Explanatory Variables

The NI is composed of indicators describing distinct physical states and referring to different population groups. Hence, despite a common unit of expression of the partial indicators – the percentage ratio - hypotheses about the determinants of NI variation have to be referenced to its components. Potential explanatory variables and their hypothetical impact on NI components are described in the subsequent section. Basic economic, political and natural factors are discussed first in Chapter 7.1.1, before health and human resource related variables are considered in Chapter 7.1.2. A summary of hypotheses is given in Chapter 7.1.3, together with a formal explanation of how the impact of explanatory variables on the NI can be assessed from the regression equations of partial indicators.

7.1.1 Basic Economic, Political and Natural Factors

The basic economic, political and natural determinants are believed to influence household food security, which is thought to be represented in the NI via the undernourishment indicator, as well as knowledge, caring capacity and health environments, which impinge on child nutrition and survival. Thus the basic factors are expected to act in varying degrees on all three components of the NI.

Macro-economic Performance, Poverty and Distribution

Macro-economic performance has already been found to be a particularly important determinant of NI variation (see Chapter 6.3.1). As far as natural resources and political decisions impinge on economic systems, their indirect impact on food security and nutrition is incorporated into this result. **National incomes** are central to the nutrition situation, because household food security, knowledge and caring capacity as well as health environments require a range of goods and services to be produced by the national economy or to be purchased on international markets. Per capita income reflects both the value of goods and services produced and the ability to satisfy demand – e.g. for food - on international markets. For countries with large agricultural sectors, per capita income includes the significant contribution of food production to overall income generated by households (Smith and Haddad 2000). Moreover, government budgets, and consequently the financial scope of policy, are limited by national resource availability. The state has to rely on taxes or other sources of finance to provide public goods like health care and education facilities, sanitation and safe water for its citizens. Per capita income comprises - by definition – 'government consumption', i.e. government spending (World Bank 2000). A drawback of the variable is its possible endogeneity, which has already been discussed in Chapter 6.3.1.

There is a strong relationship between national incomes and **poverty**, which has been 'shown by a plethora of recent studies' (Smith and Haddad 2000, p.18). The impact of poverty on household food security, use of services, care and health environments has been briefly described in Chapter 3. It would be desirable to integrate measures of income distribution and poverty into the multivariate regression. As data availability is limited and the relationship of the NI to poverty and inequality has been explored in detail in the preceding chapter, respective measures will be only briefly considered as supplementary variables in the models in Chapter 7.3.3.

Agricultural Sector and Domestic Food Production

Besides overall macro-economic output, the performance of the agricultural sector is held to be of particular importance for food security in poor countries. One NI component, the percentage of undernourished, is an indicator of household food security mainly based on national food availability. The model built to reproduce the FAO estimates of undernourishment (see Section 4.1.1) has made evident that the variation in the percentage of undernourished across countries is dominated by the variation in dietary energy supply per capita (DES), which has been pointed out by Smith 1998 also. As already mentioned, it would thus be not acceptable to use a measure of national food availability, such as DES, as an independent variable in the NI model, though dietary energy supply per capita has been shown to influence child malnutrition, and possibly acts on child mortality. Employing variables that are proximate determinants of food supply per capita seems more promising. Yet national food availability depends on production of a broad range of crops and animal products, effective demand for different foodstuffs, and on net trade balances. In order to estimate domestic production and demand as well as imports and exports of agricultural commodities, and to project dietary energy supply on an international scale, large models have been devised and applied at IFPRI and FAO (for example the 'World Food Model', see Alexandratos 1995, or IFPRI's International Model for Policy Analysis of Commodities and Trade (IMPACT), described in Rosegrant et al. 2001). These models consider, among other factors, population and income growth as driving forces of demand, and the determinants of agricultural production and international trade. Given the data and computing requirements of such huge models, their application is far beyond the scope of this study.

Yet the basic determinants described in the present section (economic performance as well as democracy, good governance, engagement in wars and regional differences in natural factors, see below) are all hypothesised to have an impact on national food availability and household food security. But finding more proximate determinants of food security which can be applied in a simple multivariate regression that does not aim at reproducing the above mentioned models turns out to be difficult. For example, yields alone do not tell very much about food availability. The dietary energy needs of a population may be likewise satisfied in a densely populated area with intense agriculture and high yields, or in a scarcely populated environment with low yields, but large areas farmed. Other potential explanatory variables are the frequency of droughts and the proportion of arable land under irrigation, which could be applied simultaneously because large-scale irrigation can offset the negative impact of drought on food production. However, these and other variables relating to agricultural productivity have a serious drawback: the importance of domestic food production for national food availability tends to decline in the course of economic development. Even the **production of major food crops per capita** is no reliable determinant of food availability, for the simple reason that more advanced countries and mineral resource exporters are able to purchase sufficient quantities of grain and other staples on international markets. In contrast, low-income countries are hardly able to buy large amounts of staple crops on the world market due to strictly limited foreign exchange reserves. This suggests the use of a variable referring to food production per capita for a subsample of poor countries only.

Democracy and Good Governance

Drèze and Sen have pointed out the crucial role of **democracy** for food security and famine prevention (Drèze and Sen 1989). Democratic governments are thought to be more responsive to the needs of all their citizens and to human rights, including the rights to food and nutrition (Smith and Haddad 2000). As has been stated in the conceptual section (see the beginning of Chapter 3), honouring other human rights, such as the right to education, basic health services, social security, gender equity, special assistance to motherhood and childhood, and freedom from discrimination, is also essential for achieving good outcomes in terms of child nutrition and survival. In addition, democratic governments are assumed to spend more of their revenues on income redistribution favouring the poor, and to be more committed to other strategies of poverty alleviation. Whereas the subjects of political decisions may differ substantially between democratic governments and authoritarian regimes, being less in favour of people's basic needs in the latter, democracy is no guarantee for **good governance**. Corruption, inefficient bureaucracies and a weak judiciary can be serious obstacles for putting well meant political decisions into practice and making a difference for food security and nutrition. As Knack emphasised: 'high levels of corruption are indicators of an environment in which redistributive rather than socially productive activities are rewarded' (Knack 1996, p. 213). Expropriation and repudiation of contracts by the government are indicative of discontinuous policies, and create a climate of insecurity that is unfavourable to private investment in agriculture and other sectors, with presumably negative consequences for food production and economic development in general.

Military Expenditure and Warfare

Spending large amounts of public resources on military purposes is hypothesised to be harmful to the nutrition situation because of the opportunity costs of **military expenditure**, which arise in times of peace as in war. Scarce governmental resources devoted to the military could be used in a socially more beneficial way in the health and education sector. However, distortion of public spending towards military expenses, accompanied by lack of public investment in agriculture, health and education, is a common feature of policy-making before and during times of war (for examples see Rwelamira and Kleynhans 1996).

Yet actual **engagement in wars** has many more detrimental consequences for food security and nutrition. Whilst wars are likely to hamper or even reverse overall economic development, there are many reasons why they may affect food security, child nutrition and survival disproportionately. Messer et al. 2001 state: 'Combatants frequently use hunger as a weapon: they use siege to cut off food supplies and productive capacities, starve opposing populations into submission, and hijack food aid intended for civilians'[2]. A variety of other effects of war contribute to large production shortfalls in agriculture and a desperate food supply situation, such as disruption of markets and the destruction of crops, livestock, roads, and land, shortage of vital inputs like fertiliser and machinery, and a lack of economic incentives for farmers (Messer et al. 2001, Guiton 2001, Rwelamira and Kleynhans 1996). Fields are sometimes not farmed because agricultural production decisions require a minimum of security about future conditions, at least until the next harvesting season. Losses to the agricultural labour force arise from killings, evictions and recruitment of the most able-bodied. Young men hide during the day for fear of forced recruitment, instead of working on the farms (reported by Feldbrügge 2000 for Angola). If people are compelled to leave their homes, the adverse circumstances of flight and migration affect their food security. They are cut off from their usual food supplies and are often hard to reach by emergency operations. Besides war-related dislocation, relative price changes, unemployment and income losses keep people from demanding food in the market (Feldbrügge 2000). The detrimental effects of violent conflict on food security do not always show up in GNP or GDP figures, because incidentally a booming war economy, fuelled by international trade in diamonds or oil, is able to disguise agricultural decline and the consequence of a desperate food supply situation for the population (see the example of Angola discussed in Chapter 6.3.1).

Regarding caring capacity as a main determinant of child nutrition and survival, it is affected by massive population displacements in violent conflicts, which often go along with household dissolution, community disintegration and breakdown of social networks. Moreover, killings deprive children of their care-takers, and together with other human rights violations they cause psychological distress. Good mental health and lack of stress have been mentioned as important resources for adequate care-giving (Engle et al. 1999). Recent studies point to severe psychiatric disorders in war refugees: a study in Cambodian refugees revealed that 68% were suffering from acute

[2] No page number can be indicated, because this is a quotation from an article in an online journal.

clinical depression, and 37% from post-traumatic stress disorder[3]. The prevalence rates found in Bhutanese refugees in Nepal and Bosnian refugees in Croatia were similar. Psychic disorders in victims of human rights violations are associated with social malfunction: 25% of Bosnian refugees in Croatia were unable to work regularly, care for their families or pursue other socially productive activities (Mollica 2000)[4]. Whereas production losses in services, industry and agriculture are captured by GNP and GDP figures, caring for family members at home – or the inability to do so due to psychosocial disorders – goes unnoticed by national accounts.

Loss of human capital and caring capacity is not only related to psychic disorders, other diseases and deaths, but is the outcome of lacking education and training in times of war also. Schools are destroyed, and the education of a whole generation is probably neglected, especially in long-lasting civil wars (Rwelamira and Kleynhans 1996). Additionally, health environments - the third main determinant of child nutrition and survival – are likely to worsen in multiple ways. Deliberate destruction of health care facilities has, for example, been reported in Liberia, Mozambique and Sierra Leone (Green 1994). Breakdown of the health care system due to lack of public funding, medical supplies and personnel can be another disastrous consequence of conflict (Criel 1998a). Furthermore, refugees and internally displaced persons in particular lack basic necessities. In refugee camps, they are frequently subject to overcrowding, poor sanitary conditions and inadequate food supplies[5]. Together with large population movements and the impairment of the health care system, the living conditions in camps facilitate the spread of infectious diseases, including HIV/AIDS (see Bucagu 2000 for a report on Rwanda).

---

[3] In non-traumatised populations, a prevalence of 10% for depressions and 8% for post-traumatic stress disorder is considered to be very high (Mollica 2000).

[4] A recent WHO report warns of a large-scale mental health crisis in Afghanistan, where 23 years of war have ravaged the psychosocial functioning of the population. Killing, executions, massive persecution, forced internal displacement, fear of landmines in mined areas, and the latest escalation of violence have had a gruesome impact not only on physical but also on mental health, causing anxiety, depression, and psychosomatic symptoms arising from psychological problems (WHO 2001).

[5] Even if the dietary energy content of food rations is sufficient, essential micronutrients are often lacking. Outbreaks of micronutrient deficiency diseases that are rarely observed in populations in their normal environment have been frequently reported from refugee camps. Baquet and van Herp 2000, for example, give an account of a pellagra epidemic in Internally Displaced People depending on World Food Programme rations around Kuito, Angola, in 1999 and 2000. They refer to a large pellagra outbreak among Mozambican refugees in Malawi in 1990 also. They note that in recent years, outbreaks of pellagra have only occurred in emergency affected populations. Pellagra can be fatal and is caused by a deficiency of niacin and/or the amino acid tryptophan in the diet.

In conclusion, wars are expected to compromise agricultural production and food security[6] as well as nutrition in multiple ways, both directly and indirectly via their impact on caring capacity, education, health care provision and health environments. Lagged effects of violent conflicts on food security and nutrition are likely to arise from the persistent legacy of landmines and the harmful long term consequences of poor schooling, psychosocial distress and epidemiological changes.

Natural factors

Natural factors like poor soil fertility influence food security directly via their impact on food production, especially in the absence of modern agricultural technology or diversified income sources. Droughts and other weather phenomena tend to affect agriculture comparatively more than other sectors of the economy. They eventually lead to food shortages, particularly in poor countries where imports are hardly able to compensate for production shortfalls. With reference to caring capacity, water scarcity as a natural factor may be limiting in rural areas: fetching water from distant sources puts a heavy strain on women's and elder children's time. Regarding health environments, they are partly shaped by geographic and climatic conditions which influence disease transmission (such as altitude, temperature, air humidity, and the like). Since comparable and reliable international data by country are mostly missing with reference to these factors, they are not considered in explicit variables. Natural factors are – besides their reflection in other explanatory variables – taken care of by **regional dummy variables** instead.

7.1.2 Health and Human Resource Related Factors

Dietary intakes that do not meet requirements for essential nutrients can be the starting point of a vicious circle affecting nutritional outcomes and survival, yet this cycle may be likewise kicked off by deficiencies in knowledge, caring capacity and health environments (see Figure 3.1-1). The following section discusses appropriate health and human resource related variables for explaining child malnutrition and mortality.

---

[6] In this context, it should be noted that variables relating to warfare may be subject to endogeneity, because *vice versa*, food insecurity can trigger off conflict (Messer et al. 2001). Yet, this may be taken into account by allowing for a time lag between the end of the fighting considered in war variables and the reference period of the NI.

Knowledge and Caring Capacity

Caring well for children as a physiologically vulnerable group with specific nutritional needs requires some minimum knowledge on nutrition and health issues. Several micro-studies have documented that **education and/or good caring practices** compensate to some extent for the adverse effect of low incomes on children's anthropometric status (Heltberg and Johannesen 2001 for Mozambique, Shori 2000 for Kenya, and Ruel et al. 1999 for Accra in Ghana). Educational achievements are assumed to be linked to urbanisation, because living in towns and cities provides better opportunities and additional incentives for investing in knowledge and skills (as urbanisation is a 'multi-purpose' variable which is also meaningful with respect to health environments, it is described in more detail below). Since women are the main care-takers of children, female education[7] is presumed to be more important for children's nutritional status than male education (Subbarao and Raney 1995, Klasen 1999, Smith and Haddad 2000, Heltberg and Johannesen 2001).

Concerning mothers' caring capacity, **discrimination against females** and **women's status** in general are hypothesised to be additional relevant factors. If women's status is low, this restricts their opportunities and freedoms, interactions with others, as well as the transmission of knowledge, and damages self-esteem (Engle et al. 1999). Women's physical status and mental health are affected by heavy discrimination against females, and if mothers are not well cared for, their children's care suffers as a consequence (Sen 1992, Ramalingaswami et al. 1996, Engle et al. 1999). Women are more likely to allocate marginal resources to their children than men, which underscores the importance of their autonomy and control over resources (Haddad et al. 1994). In societies where women's status is comparatively low, they have little say in family decisions and little control over household resources (Agarwal 1997), which limits their ability to act in their own and their children's best interest (Smith and Haddad 2000).

Besides the above mentioned variables, there are several health indicators that are directly related to caring capacity and/or care-giving: immunisation of children,

[7] It might be tempting to employ - in addition to female education - fertility as a proxy for maternal caring capacity. (The total fertility rate is defined as the number of children that would be born to a woman if she were to live to the end of her childbearing years, and bear children in accordance with prevailing age-specific fertility rates (World Bank 2000)). Caring for many children takes its toll from women's time, and short birth intervals affect their physical status. However, the problem that fertility and child mortality (as a component of the NI) are jointly endogenous (see Heltberg and Johannesen 2001) cannot be overcome, which prohibits the use of this variable.

utilisation of post- and antenatal care, and HIV/AIDS prevalence. Vaccination of children presupposes parents' willingness and ability to have their children immunised, which is likely to reflect care-takers' knowledge on health issues and commitment to their children's well-being to some extent. Post- and antenatal care usually includes advice on nutrition in pregnancy, child feeding, hygiene and handling childhood diseases, and the transmission of this knowledge fosters mothers' caring capacity (World Bank 2000). HIV/AIDS has orphaned 12 million children in Sub-Saharan Africa since the outbreak of the pandemic (UNAIDS/WHO 1998). Given the death and disease toll of AIDS in some countries, traditional kinship networks reach the limits of their capacity to provide care for orphans and the sick. Women and girls are hit hardest due to their greater social and biological vulnerability to infection and their role as care-takers for sick family members, with negative consequences for child care (FAO 2001b, Wilson 2000).

Health Care Utilisation and Health Environment

Besides individual characteristics such as nutritional and immunosystem status, utilisation of preventive care and health environments determine the risk of contracting an infectious or parasitic disease (UNICEF 1990a). Timely utilisation of curative care can break the vicious circle of infection and malnutrition at an early stage.

Immunisation protects children from infectious diseases, consecutive malnutrition and death. One-year olds are immunised against measles (necessitating one dose of the vaccine), polio, and against diphtheria, pertussis and tetanus (DPT), which is a combined vaccination requiring three sequential doses. Measles and other febrile illnesses are known to deplete Vitamin A stores (UN ACC/SCN 1997). Vitamin A deficiency impairs child growth and endangers survival (MI/UNICEF/Tulane University 1998). Respiratory diseases such as diphtheria and pertussis, which is better known as 'whooping cough', may be fatal and are likely to affect nutritional status. Furthermore, the immunisation coverage is in a sense indicative of the effectiveness of the health care system.

Adequate **health care for women during pregnancy, childbirth and after delivery** (i.e. midwifery, post- and antenatal care) improves and preserves mothers' health and reduces maternal and infant mortality (World Bank 2000). In addition, training health staff for ante- and postnatal care reflects public responsiveness to mothers' and children's needs. The variable may thus give an indication of women's and children's societal status.

Access to health care is presumed to be a weaker (though possibly significant) explanatory variable of child nutrition and survival than actual utilisation of services. Generally access indicators capture the supply of services, but do not reflect rates of use or benefits. This holds in particular for access to health care: first, physical access is no guarantee of financial accessibility by all strata of the population, which may be prevented by official user fees or under-the-table payments demanded at the point of use (Criel 1998b, Gilson 1998). Second, essential drugs and other medical supplies are frequently unavailable at health facilities in poor developing countries, especially in Sub-Saharan Africa (Criel 1998b, Musau 1999). Third, actual and/or perceived quality of care has been found to be so low in some public health facilities that people are not willing to utilise the services offered (see Batusa 1999 for an illustrative example). Clearly, a decent quality of care is relevant for health outcomes.

The risk of contracting contagious diseases depends not only on the utilisation of preventive services, but also on the health environment. Lack of **safe drinking water and basic sanitation** are main reasons for the high prevalence of diseases transmitted by faeces in developing countries (World Bank 2000). Diarrhoea belongs to this type of diseases, which can push children rapidly into the vicious circle of infection, nutrient loss and malnutrition, possibly with deadly consequences. Frequent episodes of diarrhoea require relatively higher levels of household food security to counteract child malnutrition (Haddad et al. 1998).

Regarding health environments and the outreach of public services in general, **urbanisation** is considered as a proxy variable[8]. Health care facilities tend to be concentrated in urban areas (World Bank 2000). The quality of health care is assumed to be better in urban than in rural areas. For evident reasons, it is much more cost-effective (and for some types of facilities, e.g. in the case of sewerage and piped water, more imperative) to provide basic public services to people in a densely populated urban area than to serve a widely dispersed rural population. For Mozambique, Heltberg and Johannesen found that – controlling for parental education and income - children in urban areas had significantly better health outcomes and achieved more years of

---

[8] Since rural poverty and food insecurity have been cited as causes of accelerated rural-urban-migration (von Braun 1996), joint endogeneity of the urbanisation variable and the percentage of undernourished may be a problem. Yet, despite the ongoing shift of food insecurity and malnutrition to the cities, the majority of the poor and food insecure still reside in rural areas (Garrett 2001), which renders a positive correlation of the two variables unlikely.

schooling than rural children, which presumably reflects differences in availability of public services (Heltberg and Johannesen 2001).

Besides the physical environment and access to public services, the pattern of prevailing communicable diseases in a population plays a role for infection risks, child nutrition and survival. For example, full-blown AIDS is usually accompanied by progressive wasting, and favours the spread of other contagious diseases like tuberculosis. An infant infected with HIV has no survival chances without expensive medical treatment, which is unaffordable to most people in developing countries. The mother-to-child transmission rate being between 25% and 35%, HIV is contributing substantially to increasing infant and child mortality rates in the Sub-Saharan countries worst affected by **HIV/AIDS**[9] (UNAIDS/WHO 1998).

7.1.3 Summary of Hypotheses

Hypotheses on explanatory factors of NI components have been outlined in the preceding subchapters. Prior to an overview of hypotheses in tabular form, the potential impact of explanatory variables on the NI is formally derived in the following section. An estimation model of the NI is thought to be an aggregate of three single indicator models. In abbreviated vector notation (see Box A 7.1.3-1 in the Appendix for the full forms), the regression equations of NI components are expressed as

$$(7.1) \quad y_{PUN} = \mathbf{X}\beta_{PUN} + u_{PUN}$$

$$(7.2) \quad y_{CUW} = \mathbf{X}\beta_{CUW} + u_{CUW}$$

$$(7.3) \quad y_{CM} = \mathbf{X}\beta_{CM} + u_{CM}$$

For N countries in the data set and k explanatory variables, y and u are N x 1 vectors of dependent variables and error terms, respectively, and β denominates a (k+1) x 1 vector containing the parameters, i.e. the intercept and the coefficients of the k independent variables. **X** is an N x (k+1) data matrix, with N values of 1 in the first column for multiplication with the intercept β_0, and the data values of k common explanatory variables in the subsequent columns. The subscripts of y, u and β indicate to which dependent variables the vectors in the regressions pertain: PUN (percentage of the Population Undernourished), CUW (percentage of Children Underweight), or CM

[9] In 1997, HIV/AIDS prevalence in adults exceeded 20% in Namibia, Zambia, South Africa and Lesotho, and was over 25% in Botswana, Zimbabwe and Swaziland (UNAIDS/WHO 1998).

(Child Mortality, i. e. percentage of children dying before age five). Accordingly, the regression equation for the NI is

$$(7.4)\quad y_{\mathrm{NI}} = \mathbf{X}\beta_{\mathrm{NI}} + u_{\mathrm{NI}}$$

If the NI calculation Formula (4.1) in Chapter 4 is rewritten in vector notation for a set of N equations pertaining to N Nutrition Index values, this is denoted as follows

$$(7.5)\quad y_{\mathrm{NI}} = c - \left(0.33 y_{\mathrm{PUN}} + 0.32 y_{\mathrm{CUW}} + 0.35 y_{\mathrm{CM}}\right)$$

when $c = \begin{bmatrix} 100 \\ \vdots \\ 100 \end{bmatrix}$, N x 1 vector containing N times the value 100

According to the NI calculation formula, the aggregated single indicators are subtracted from 100, which is represented by vector c in Equation (7.5). The right hand sides of Equations (7.1) to (7.3) can be likewise aggregated by weighting and adding up. This means that Equation (7.5) is equivalent to

$$(7.6)\quad y_{\mathrm{NI}} = c - \mathbf{X}\left(0.33\beta_{\mathrm{PUN}} + 0.32\beta_{\mathrm{CUW}} + 0.35\beta_{\mathrm{CM}}\right) - \left(0.33u_{\mathrm{PUN}} + 0.32u_{\mathrm{CUW}} + 0.35u_{\mathrm{CM}}\right).$$

If the error terms in the regressions of NI components are independently distributed, which is the case if they are normally distributed and uncorrelated[10] (Gujarati 1995), the intercept of the NI model, see Equation (7.4), is directly derived from the above as:

$$(7.7)\quad \beta_{\mathrm{NI0}} = 100 - \left(0.33\beta_{\mathrm{PUN0}} + 0.32\beta_{\mathrm{CUW0}} + 0.35\beta_{\mathrm{CM0}}\right)$$

Again, supposing independently distributed error terms, the coefficients of explanatory variables in the NI model are equivalent to the weighted sums of the coefficients of the single indicator models, multiplied by -1:

---

[10] The error term comprises measurement error in the dependent variable as well as the impact of omitted and unobservable variables on the dependent variable (Gujarati 1995). In an Ordinary Least Squares (OLS) regression, the error term includes country fixed effects that are not explained by or correlated to the independent variables. It is unlikely that the error terms of the partial indicator regressions in (7.1) to (7.3) are correlated to each other, unless there is substantial omitted variable bias in the models, or there are large and correlated measurement errors in dependent variables.

$$(7.8)\quad \begin{bmatrix} \beta_{\mathrm{NI1}} \\ \beta_{\mathrm{NI2}} \\ \vdots \\ \beta_{\mathrm{NIk}} \end{bmatrix} = -\begin{bmatrix} 0.33\beta_{\mathrm{PUN1}} + 0.32\beta_{\mathrm{CUW1}} + 0.35\beta_{\mathrm{CM1}} \\ 0.33\beta_{\mathrm{PUN2}} + 0.32\beta_{\mathrm{CUW2}} + 0.35\beta_{\mathrm{CM2}} \\ \vdots \\ 0.33\beta_{\mathrm{PUNk}} + 0.32\beta_{\mathrm{CUWk}} + 0.35\beta_{\mathrm{CMk}} \end{bmatrix}$$

For a meaningful interpretation of the parameters in the NI model and in order to trace back the significance of explanatory variables to their impact on NI components, the single indicator models have to be considered.

In the same manner like the coefficients in Equation (7.8), the error term of the NI model can be expressed as the aggregate of error terms of NI component models. In case the error terms of the single indicator regressions are independently distributed, the impact of error will decline in the NI model as compared to the role of error in the three component models on the average. This means, in other words and more precisely, that the variance of the error term in the NI model, $\sigma^2_{u_{\mathrm{NI}}}$, will be smaller than the weighted sum of the variances of error terms[11] in the single indicator regressions:

$$\begin{aligned}
(7.9)\quad & \sigma^2_{u_{\mathrm{NI}}} && \le 0.33\sigma^2_{u_{\mathrm{PUN}}} + 0.32\sigma^2_{u_{\mathrm{CUW}}} + 0.35\sigma^2_{u_{\mathrm{CM}}} \\
\Leftrightarrow\quad & \sigma^2_{-(0.33u_{\mathrm{PUN}} + 0.32u_{\mathrm{CUW}} + 0.35u_{\mathrm{CM}})} && \le 0.33\sigma^2_{u_{\mathrm{PUN}}} + 0.32\sigma^2_{u_{\mathrm{CUW}}} + 0.35\sigma^2_{u_{\mathrm{CM}}} \\
\Leftrightarrow\quad & (-0.33)^2\sigma^2_{u_{\mathrm{PUN}}} + (-0.32)^2\sigma^2_{u_{\mathrm{CUW}}} + (-0.35)^2\sigma^2_{u_{\mathrm{CM}}} && \le 0.33\sigma^2_{u_{\mathrm{PUN}}} + 0.32\sigma^2_{u_{\mathrm{CUW}}} + 0.35\sigma^2_{u_{\mathrm{CM}}} \\
\Leftrightarrow\quad & 0 && \le 0.2211\sigma^2_{u_{\mathrm{PUN}}} + 0.2178\sigma^2_{u_{\mathrm{CUW}}} + 0.2275\sigma^2_{u_{\mathrm{CM}}}
\end{aligned}$$

The transformation of the inequation confirms that the left hand side (variance of error term in NI model) is smaller than the right hand side (aggregated variances of error terms of single indicator models)[12]. Both sides of the inequation would only be equal in cases where the variances of error terms in the single indicator models were all equivalent to zero. This would require that all predicted values in the NI component models matched actual values perfectly, which is virtually impossible.

The fact that the variance of the error term in the NI model, $\sigma^2_{u_{\mathrm{NI}}}$, will decrease as compared to the term $0.33\sigma^2_{u_{\mathrm{PUN}}} + 0.32\sigma^2_{u_{\mathrm{CUW}}} + 0.35\sigma^2_{u_{\mathrm{CM}}}$ (if the error terms of the compo-

---

[11] If X and Y are independent random variables, a and b are constants and $Z = aX + bY$, then $\sigma^2_Z = a^2\sigma^2_X + b^2\sigma^2_Y$ (adapted from Gujarati 1995).

[12] Because the variances are squared figures, which cannot turn negative, it is evident that the bottom line inequation in (7.9) is true, and so are the preceding expressions.

nent regressions are normally distributed and uncorrelated) still does not prove that the explanatory power of the NI model will increase as compared to the weighted average explanatory power of the NI component models. Changes in the variance of NI values vis-à-vis the variances of NI components have to be taken into account. Explanatory power, also called 'goodness of fit', is indicated by the coefficient of determination, R-squared[13]. R-squared is defined as follows (adapted from Gujarati 1995):

$$(7.10)\quad R^2 = \frac{\text{Explained Variance of y}}{\text{Total Variance of y}} = \frac{\text{Total Variance of y} - \text{Unexplained Variance of y}}{\text{Total Variance of y}}$$

$$\Leftrightarrow \quad R^2 = 1 - \frac{\sigma_{\hat{u}}^2}{\sigma_y^2}.$$

Note that the variance of $\hat{u}$ is used instead of the variance of u in Equation (7.10), because R-squared refers not to the error terms in the models (u), but is derived from the estimated error terms ($\hat{u}$), i.e. the residuals of the regression. The assumption that the explanatory power of the NI model is relatively higher than (or at least equal to) the coefficients of determination of single indicator regressions, aggregated by NI weights, can be denoted as

$$(7.11)\quad R^2_{\mathrm{NI}} \geq 0.33\ R^2_{\mathrm{PUN}} + 0.32\ R^2_{\mathrm{CUW}} + 0.35\ R^2_{\mathrm{CM}}$$

$$\Leftrightarrow \quad 1 - \frac{\sigma^2_{\hat{u}_{\mathrm{NI}}}}{\sigma^2_{y_{\mathrm{NI}}}} \geq 1 - 0.33\frac{\sigma^2_{\hat{u}_{\mathrm{PUN}}}}{\sigma^2_{y_{\mathrm{PUN}}}} - 0.32\frac{\sigma^2_{\hat{u}_{\mathrm{CUW}}}}{\sigma^2_{y_{\mathrm{CUW}}}} - 0.35\frac{\sigma^2_{\hat{u}_{\mathrm{CM}}}}{\sigma^2_{y_{\mathrm{CM}}}}$$

$$\Leftrightarrow \quad \frac{\sigma^2_{\hat{u}_{\mathrm{NI}}}}{\sigma^2_{y_{\mathrm{NI}}}} \leq 0.33\frac{\sigma^2_{\hat{u}_{\mathrm{PUN}}}}{\sigma^2_{y_{\mathrm{PUN}}}} + 0.32\frac{\sigma^2_{\hat{u}_{\mathrm{CUW}}}}{\sigma^2_{y_{\mathrm{CUW}}}} + 0.35\frac{\sigma^2_{\hat{u}_{\mathrm{CM}}}}{\sigma^2_{y_{\mathrm{CM}}}}$$

The inequation sign is reversed in the last transformation step due to multiplying by –1. If the variances of error terms in the NI and the single indicator regressions equalled the variances of the corresponding dependent variables, all coefficients of determination would be zero, and so would be the left hand side and the right hand side of the inequation. However, this would only occur in cases where the common independent variables were not able to explain any part of the variation of the three

[13] Adjusted R-squared is not considered here for the sake of simplicity and because the number of independent variables – for which the adjustment of the coefficient of determination corrects - is the same for the NI model and the single indicator regressions.

variables PUN, CUW, and CM, which is very unlikely[14]. If it actually came true, the error terms in the regressions of NI components would equal the dependent variables and, therefore, be inevitably correlated to each other, since NI components themselves are correlated. In this case the following substitution of $\sigma^2_{\hat{u}_{\mathrm{NI}}}$ would not be correct. Yet under the assumption of independently distributed residuals, see the left hand sides of Inequation (7.9) for comparison,

$$(7.12)\ \sigma^2_{\hat{u}_{\mathrm{NI}}} \quad = 0.1089\sigma^2_{\hat{u}_{\mathrm{PUN}}} + 0.1024\sigma^2_{\hat{u}_{\mathrm{CUW}}} + 0.1225\sigma^2_{\hat{u}_{\mathrm{CM}}}.$$

In addition, the known empirical variances of the NI and its components (tabulated in Chapter 4.3) are inserted in Inequation (7.11), resulting in

$$(7.13)\ 0 \quad \leq 0.00019\sigma^2_{\hat{u}_{\mathrm{PUN}}} + 0.00088\sigma^2_{\hat{u}_{\mathrm{CUW}}} + 0.00584\sigma^2_{\hat{u}_{\mathrm{CUW}}}$$

The numbers are rounded to the fifth decimal place. The right hand side of the inequation is larger than the left hand side, except for the extremely improbable case that the error terms of the single indicator regressions have zero variance (which would require that all residuals are zero). This confirms the assumption of relative gains in explanatory power with respect to the NI model and for the given data set, if error terms are in fact independently distributed[15].

Finally, Table 7.1.3-1 gives an overview of the explanatory variables previously discussed and their hypothesised impact on NI components and the NI. Since the NI increases with improvements in food security and nutrition, factors diminishing undernourishment, child malnutrition and mortality - showing negative signs in respective regressions - have a positive effect on the NI, indicated by positive signs of coeffi-

---

[14] In fact the investigation in Chapter 6.3.1 has already shown that GNP per capita is a very powerful explanatory variable of the NI as an aggregate.

[15] The result can be explained as follows: the variance of the NI is smaller than the weighted average of variances of PUN, CUW and CM, but not as much smaller as the variance of the residual of the NI model compared to the aggregate variances of residuals of NI component models. Consequently, this leads to gains in R-squared, see Equation (7.10). The deeper reason is that NI components are positively correlated, which means that their positive covariances have to be added when the variance of the NI is derived from the variances of its partial indicators. In contrast, the error terms of single indicator regressions can reasonably be assumed to be independently distributed, with covariances being equivalent to zero.

cients in the NI model (see Equation (7.8)). Accordingly, positive signs in the NI component models correspond to a negative sign in the regression of the NI.

Table 7.1.3-1 Expected signs of coefficients in regressions of the NI and its components on explanatory variables

EXPLANATORY VARIABLES	DEPENDENT VARIABLES			
	NI COMPONENTS			NI
	PUN	CUW	CM	
1. Basic Economic and Political Determinants				
Macro-economic performance	-	-	-	+
Poverty/Inequality	+	+	+	-
Domestic food production (in poor countries only)	-	- (?)	- (?)	+
Democracy	-	-	-	+
Good governance	-	-	-	+
Military expenditure	+	+	+	-
War	+	+	+	-
2. Health and Human Resource Related Determinants				
(Female) education	0	-	-	+
Discrimination against women	0	+	+	-
Immunisation of children	0	-	-	+
Midwifery, ante- and postnatal care	0	-	-	+
Health care/Safe water/Sanitation	0	-	-	+
Urbanisation	0	-	-	+
HIV/AIDS	0	+ (?)	+	-

Source author's own presentation

The effect of some explanatory variables on NI components is hypothesised to be weak, which is depicted by a question mark in brackets. Furthermore, discrimination against females and HIV/AIDS might have a negative impact on food production and dietary energy supply (as has been argued by Quisumbing et al. 1995 for discrimination against women and by Wilson 2000 and FAO 2001b for HIV/AIDS). Thus, these variables might tend to raise the proportion of undernourished in the population, which is not indicated in the table. Before turning to test empirical hypotheses, attention is devoted to operational measures of variables and data sources.

7.2 Operational Measures of Variables and Data Sources

Not all potential determinants of NI variation lend themselves readily to numerical expression. Moreover, data availability is limiting in many instances. In the present subchapter, indicators employed as explanatory variables are discussed, and data sources and transformations are described. The dependent variables - the NI and its components - are briefly considered, because predicted values have been used to supplement the data set, which might bias estimation results.

7.2.1 Dependent Variables

To obtain a broad set of cross country data for the multivariate regression, NI 1997 figures are supplemented by NI 1992 values. Since data sources, refinements and aggregation of NI components as well as reference years have been detailed in Chapter 4, their description is not repeated here. Yet the fact that the NI values of some countries are based on estimated values of partial indicators has to be considered for the purpose of modelling. For some of the countries concerned, data are lacking in many fields, and they drop out of the analysis anyway due to unavailability of basic macro-economic statistics[16]. For 95 nations, NI values for 1997 and/or 1992 as well as concurrent data on national incomes are available (see Chapter 6.3.1 for a list of the nine countries that could not be included). This set of countries is subsequently referred to as the 'basic data set' or 'basic sample'.

Ten of the 95 countries have NI values comprising the author's own **estimates of undernourishment**, which have been predicted by the model devised in Chapter 4.1.1. It may be argued that the model reproduces the FAO estimation process very well, see Table 4.1.1-1 and Figure 4.1.1-2. But what is more, seven of the ten cases are countries in transition, allowing for a comparison of predicted undernourishment rates with the figures published in FAO 2000a for this region. The estimates generated by the model are found to be well in line with the FAO figures released for these countries[17]. No corresponding check can be made for the remaining three countries – Comoros, Fiji, and South Africa. However, it should be stressed that data from anthropometric surveys but no estimated values of child malnutrition have been included in the NI for all ten countries under consideration. The anthropometric data might possibly balance estimation errors in the undernourishment component. Only four of the 95 countries in the basic data set have NI values that include **estimates of underweight in**

[16] In the absence of data on national income, which has already been found to be of crucial importance for the nutrition situation as measured by the NI (see Chapter 6.3.1), the intended investigation makes little sense. For this reason, countries without information on GNP per capita are not included in the multivariate regression, and are not considered any further in the present chapter.

[17] If dietary energy supply per capita of 1996-98 (three-year average, published in FAO 2000a) and its logarithm are applied as independent variables in the estimation model, deviations of predicted values from FAO figures for undernourishment (1996-98) are found to be between –3.3 and +2.3 percentage points (the extreme values of the residuals pertain to Azerbaijan and Uzbekistan). The mean value of residuals amounts to –0.39 percentage points.

children taken from UN ACC/SCN 1993: Gabon, Guinea, Saudi Arabia and Swaziland. For Guinea and Swaziland, residuals generated by the ACC/SCN model (-4.2 and –1.2, respectively, relating to actual prevalence rates of 23.4 and 9.7) indicate good estimates for these countries (see UN ACC/SCN 1993). Solely for Gabon and Saudi Arabia, a comparison of actual and predicted prevalence rates of underweight in children is not possible due to the absence of nationally representative surveys.

In conclusion, it is very unlikely that retaining the 14 countries with NI values partly based on predicted values in the data set (or the even smaller number of five countries altogether for which no direct validation of estimates could be performed) might cause significantly biased parameter estimates in the regression. Therefore, no countries are omitted from the basic sample due to the inclusion of estimates in the NI.

7.2.2 Independent Variables

Operational measures of independent variables and the data used are discussed in the same sequence in which the respective explanatory factors have been presented in Chapter 7.1.1. If not stated otherwise, data for the independent variables under discussion are available for all 95 countries in the basic data set.

Macro-economic Performance

A decision has to be made whether Gross Domestic Product (GDP) or Gross National Product (GNP) per capita is to represent a country's economic performance in the regression. GDP measures the total output of goods and services for final use generated within the domestic territory of a given country, regardless of the allocation to domestic and foreign claims. This means that the goods and services produced by migrant workers from abroad or foreign firms are also considered. In contrast, GNP includes the production of goods and services by country nationals, expressed in monetary terms, no matter if they live in their home country or abroad. For most countries, the difference between GDP and GNP is quite small (World Bank 2000). There are arguments in favour of both indicators when it comes to selecting one as an explanatory variable in the model. On the one hand, resident and non-resident firms pay taxes which can be used by the state to provide public goods of relevance for food security and nutrition. This would suggest choosing GDP per capita as a variable. On the other hand, in many low-income countries remittances from family members living and working abroad constitute a large share of the income for poor households, thereby contributing to purchases of food and other basic necessities. Remittances are not part of GDP, but they are considered in GNP (at least as far as possible), which speaks in favour of the latter. Given the pros and cons, GNP per capita has finally been chosen because of the inclusion of remittances and for the reason that the contribution of economic production to tax income and the provision of public goods, which is probably better reflected by GDP, can also be proxied by variables directly relating to public services. GNP per capita is expressed in Purchasing Power Parity dollars (PPP$) and in real terms, with 1995 as the base year. For the food security and nutrition situation, the purchasing power of people on domestic markets is far more important than their ability to buy on international markets, which justifies the preference of **GNP per capita in constant 1995 PPP$** to GNP per capita in US$, calculated from official exchange rates. The same data that have been used in the preceding chapter for correlation analyses (taken from World Bank 2000 and adapted to NI reference years) are used here. The careful matching with NI reference years is retained due to the predominant role of the GNP per capita variable on the one hand, which has become evident in Chapter 6.3.1 in a univariate regression, and because this indicator is eventually subject to rapid shifts. Taking into account the functional relation of NI and GNP per capita identified in Chapter 6.3.1 (see the scatterplot in Figure 6.3.1-1), the variable is applied in logarithmic transformation.

Poverty and Distribution

Among indicators of economic inequality, the **Gini coefficient** and the **ratio of the top quintile share to the bottom quintile share of income or consumption (R20)** are selected, see Chapter 6.2.1 for a description of the data. Measures of distribution are available for only 71 countries in the basic sample. Poverty indicators (**percentage of the population below 1$ and below 2$ a day** as well as **poverty gap at 1$ and 2$ a day**) and respective data are detailed in Chapter 6.2.2. Seven countries with NI 1997, but no poverty data later than 1993 (for which NI 1997 has been replaced by NI 1992 in the rank correlation analyses in Chapter 6.2.2) have been excluded. Therefore, only 52 countries with information on poverty are found in the basic data set.

Domestic Staple Food Production

Since starchy staples account for the largest share of dietary energy supply in low-income countries - roughly 60-80%, as already mentioned in Chapter 6.3.1 - domestic production of this food category is most important for dietary energy intakes in these countries. Attention is, therefore, focussed on the production of cereals, roots and tubers. Roots and tubers should not be neglected, because in some countries the diet is traditionally based on crops like cassava and yams and much less on cereals (a prime example is Ghana in West Africa). The basic data for the variable are obtained from the FAOSTAT database (FAO 2000b). The production of cereals (excluding cereals used for beer production) on the one hand and roots and tubers (dry equivalent) on the other are given as aggregate figures in 1000 tons[18]. Together with population data, these figures allow the calculation of **cereal and roots and tubers production in kg per capita**. Average values are formed for the years 1990-92 and 1995-97 and matched with NI 1992 and NI 1997, respectively.

Democracy

A wide array of institutional variables has been applied in the research on economic growth. The **Gastil indices** released by Freedomhouse, for example, attempt to quantify the degree of **political freedom** and **civil liberties**[19] in a country. They are seven-point country-level indices, with higher values indicating less freedom, and are based on subjective assessment. As political freedom and civil liberties can be quickly wiped out, e.g., in the event of a coup d'état, three-year averages are formed for 1990-1992 and 1995-97 and matched with countries' NI values of 1992 or 1997. Data are obtained from the ratings published annually by Freedomhouse (Freedomhouse 2000). Possible lagged effects are not considered.

Good Governance

Another category of institutional indicators is compiled in the ICRG (International Country Risk Guide) to assess the degree of country risk faced by multinationals. Though the measures are subjective and tailored to the information needs of international investors, they reflect to some extent the quality of governance and the institutional environment. The aggregate **ICRG index** comprises five separate **indices for 'Quality of the Bureaucracy', 'Corruption in Government', 'Rule of Law', 'Expropriation Risk', 'Repudiation of Contracts by Government'**[20]. The first three

---

[18] The caloric values of the different foodstuffs summarised in each category are roughly equal, as roots and tubers are given in dry equivalents. Hence, weighting the amounts of specific cereals, roots and tubers according to their caloric values is not necessary.

[19] The natural average of these two indices has been employed as variable of democracy in a model explaining child malnutrition, and has proved significant (Smith and Haddad 2000).

[20] Excerpts from variable descriptions in the International Country Risk Guide can be found in Box A 7.2.2-1 in the Appendix.

variables listed are measured on a 0-6 scale, the subsequent two indicators of risk on a 0-10 scale. After transforming the first three measures to a 0-10 scale, these indices can be aggregated by simple addition to form the ICRG index, as has been done by Knack and Keefer (1995) for their model of economic growth[21]. Since changes in the institutional environment can occur rapidly, and ICRG index data are available for each year in the nineties, an average value of the years 1990-92 is formed for countries with NI 1992, and a mean value for 1995-97, if NI 1997 is used. Data are taken from the IRIS-3 database constructed by Stephen Knack for the IRIS Center, University of Maryland, from monthly ICRG data provided by The PRS (Political Risk Services) Group (Knack 1999). A disadvantage of the ICRG measures lies in the fact that they are available for only 75 countries in the basic sample.

Military spending

Data on military expenditure are published by the World Bank as a percentage of central government expenditure and as a percentage of GNP. Since GNP per capita is used in the model, **military spending expressed as a proportion of GNP** is most suitable, because the impact of military expenditure can then be displayed against the GNP variable. Data for other than NATO countries generally cover expenditures of the ministry of defence, whereas expenditures on public order and safety are classified separately and not included (World Bank 2000). For most countries with GNP data, information on the amount of military expenditure is available throughout the nineties. The natural average of military expenditure data for the years 1990-92 and 1995-97 was formed for countries with NI 1992 and NI 1997, respectively. For a few countries, data were not available for all three years of the period in question. In these cases, the average level of military spending for only two years was calculated, or, if data were available for no more than one year in the three-year period, this figure was assumed to be representative for the triennium. An exception was made for Haiti with NI 1997, for which no data on military expenditure were available from 1995-97. However, data from the years 1990-1993 indicate a stable level of military spending during that time. Therefore, the figure relating to 1993 was used for Haiti. In the end, only one country (Comoros) had to be dropped from the basic sample due to complete lack of data on military expenditure in the nineties.

[21] Weighting by factor scores generated from factor analysis yields a correlation of 0.99 with the additive index (Knack and Keefer 1995).

Engagement in Wars

As has been outlined in Chapter 7.1.1, wars can have long-lasting negative effects on food security and nutrition, which cannot be eliminated within a short time after peace has been restored. For this reason and because data for NI calculation were drawn from a period of six years, wars that ended some time before the actual reference year of the NI (1997 or 1992) should be taken into account in the design of a **war dummy** variable (which constitutes the simplest option to represent the potential effects of war in the regression). However, *a priori* predications about lagged effects and the time spans to be considered are hardly possible. Therefore, two war dummy variables with different time horizons were constructed. The first variable considers wars that had started, were ongoing, or had ended within the four years preceding the reference year of the country's latest NI, or in the reference year itself. The relevant time periods are, therefore, 1993-97 for countries with NI 1997, and 1988-92 for countries with NI 1992 only. The second variable takes into account warfare occurring within a ten-year period: 1988-97 for countries with NI 1997 and 1983-92 for countries with NI 1992 only[22]. The first dummy variable assigned the value of one to all countries found at war at some point within the five-year period, the second one assumed the value of one if a country was engaged in a war within part or all of the ten-year period, and both variables were set to zero for non-war countries.

Assessing the severity of wars is more difficult than simply distinguishing between war and non-war countries. Data on the number of deaths, for example, are not always available and mostly refer to military personnel only (HIIK 2000). Yet information on **war duration** at a chosen point in time is broadly available and might give at least some indication of the severity and the disastrous cumulative impact of wars. Thus, two additional variables were designed which indicate war duration for the countries classified as war countries in the five-year period and the ten-year period, respectively, assigning the value of zero to non-war countries. All years under warfare within the reference period of five or ten years are summed up, irrespective of interruptions by times of peace. Countries afflicted with wars lasting only a few months were assigned the variable value of one, like countries in which wars were carried on for one full year up to less than two years. If a country was engaged in two different

---

[22] In addition to the five-year variable, the ten-year variable contains information on warfare in Cambodia during 1978-1991, the civil war in El Salvador in 1981-1992, the first Gulf war between Iran and Iraq, and wars in Nicaragua, Chad, Lebanon and Kuwait. Iraq and Kuwait, however, are not in the basic data set due to lack of GNP data.

wars within the same year, no double counting was done but the respective year was only taken into account once. For wars that ended within the reference period, but began at some point in time before, their full duration was considered[23]. Data were taken from the KOSIMO database of the Heidelberg Institute for International Conflict Research (HIIK 2000).

Regional Dummy Variables

Regional dummy variables, assigning the value of one to countries in a specific group and zero to all other countries, were built for six out of seven regions: **the Caribbean, Eastern Europe & the former Soviet Union, Middle East & North Africa, South Asia, East Asia & Pacific**, and **Sub-Saharan Africa**[24]. All other countries in the data set are in Latin America. No dummy variable was constructed for this region, following the rule that the number of dummies has to be one less than the number of categories of the qualitative variable (which is 'region', in this case), see Gujarati 1995. Latin America thus serves for comparison in the regression and was chosen as the base group, because it neither did extraordinarily well nor particularly badly in past models of child malnutrition or child mortality. This means that, in contrast to South Asia for example, the performance of Latin American countries with respect to child nutrition and survival was largely explained by the continuous variables in these models.

---

[23] Some examples may help to clarify the process: In Yemen the 70-days-war was waged in 1994. NI 1997 is available and used for Yemen in the data set, and so the reference period for the five-year variable is 1992-97, and the relevant time span for the ten-year variable is 1988-1997. The conflict year is part of both reference periods and, consequently, both war duration variables are set to one – as stated above, the variables are not differentiated any further according to days or months of conflict duration. In Lebanon, there was a war starting in 1989 and ending in 1990 (previous episodes of violent conflict are not considered because they did not extend to the reference periods). Since NI 1997 is used for Lebanon, the time periods taken into account are again 1992-1997 and 1988-1997. Lebanon is assigned a war duration of zero according to the five-year variable, because the war ended before 1992, and the value of one for war duration according to the ten-year variable. The value of one is chosen although the war in 1989-1990 may have lasted more than one year (or possibly less than one year), because its duration surely did not reach two years (no precise dates for the beginning and end of conflict are given in the data source). Eritrea had been engaged in a civil war from 1967 to 1993. The latest NI for Eritrea refers to 1997, and because the war extended into both time periods under consideration for the variable, they are both assigned the value of 26. Thus the full duration of uninterrupted fighting until the end of the conflict in 1993 is taken into account.

[24] See Table A 7.2.2-1 in the Appendix for the classification of countries in the basic data set by region.

Education

The **total literacy rate** - usually defined as the proportion of people aged 15 and above who can read and write - is claimed to be a variable that reflects basic education in the adult population. However, this indicator has serious drawbacks, particularly deficiencies in international comparability (see the comments in Chapter 2.1.2), which have to be kept in mind. As literacy is a stock variable that develops slowly over time, no average values for several years are formed, but data for 1992 and 1997 are taken from the Human Development Reports 1994 and 1999 (UNDP 1994 and UNDP 1997) and matched with NI reference years. Figures on literacy rates published by UNDP are broadly available (for 94 out of 95 countries in the basic data set), because they were partly complemented by estimates.

Since female education is held to be more important for preventing child malnutrition and mortality than male education, data on the **female literacy rate** are calculated from figures on female illiteracy published by the World Bank (World Bank 2000), assuming that all women that have not been classified as illiterate are able to read and write. Yet data on female literacy rates are not as broadly available as data on total literacy rates for whole populations– for 88 out of 95 countries in the basic data set only - and the same critique with respect to insufficient international comparability applies. Another eligible educational indicator is the **female gross secondary school enrolment rate**, which Haddad and Smith (2000) have used in their explanatory model of child malnutrition. However, enrolment rates are flow variables relating to girls and do not properly reflect the education received by all women of reproductive age, which is most relevant for child care (though high correlation of present secondary enrolment with enrolment rates in the past is likely). Furthermore, net enrolment rates would be preferable to gross enrolment rates, because they are not blurred by inefficiencies of the educational system such as a high proportion of students repeating classes[25]. So, female gross secondary enrolment rates are chosen instead of net enrolment rates, because data availability is too scanty for the latter. For countries with NI 1992, the most recent figures for the years 1990 to 1992 are used, and the latest information from the years 1993 to 1997 has been considered for countries with NI 1997. Respective data are available for 83 of the 95 countries in the basic data set.

---

[25] Net enrolment is the ratio of the number of children of official school age (as defined by the national education system) who are enrolled in school to the population of the corresponding official school age, whereas gross enrolment is the ratio of total enrolment, regardless of age, to the population of the age group that officially corresponds to the level of education shown (World Bank 2000).

Discrimination against Women

Discrimination against women or, more generally, 'women's relative status' is not easy to capture. Smith and Haddad (2000) have employed the **ratio of female to male life expectancy** as a proxy for women's relative status to explain child malnutrition. Yet there are many other factors influencing gender differentials in mortality than gender inequity and discrimination, e.g. wars, different biological susceptibility of the sexes to HIV/AIDS and other diseases, fertility levels and related maternal mortality, the impact of gender roles on smoking, drinking and other self-damaging activities (Rizo 1996, Santow 1996, Waldron 1985). Nevertheless, the ratio of female to male life expectancy may serve as a very crude measure of women's well-being relative to men's in a society. Since life expectancy is subject to slow changes, data for the years 1992 and 1997 are considered representative for the reference periods of NI 1992 and NI 1997, respectively, and are taken from World Bank 2000 for both males and females in order to calculate the ratio.

Compared to the ratio of female-to-male life expectancy, differentials in child mortality are less blurred by factors not related to discrimination that confound mortality rates of adult males and females (Waldron 1987). Besides inherent biological differences, they are likely to reflect the relative valuation of boys and girls, which is supposed to be closely linked to the relative status of men and women (compare Bardhan 1982, Sen 1990). Therefore, the **ratio of male-to-female under-five mortality** is used as an alternative proxy for the societal status of women. Under-five mortality data disaggregated by gender are available from WHO (1999) for 1998, which does not allow for close matching with NI reference years. However, the figures may be approximately representative for earlier years as well.

Health Care Utilisation

The most recent data on **immunisation against measles and DPT** out of the periods 1990-92 and 1995-97 have been taken from World Bank (2000) and have been matched with the NI reference years 1992 and 1997, respectively, for countries in the data set. As immunisation rates are subject to rather slow changes, no average values were calculated for three-year periods.

Another variable referring to utilisation of health services is the **percentage of births attended by health staff**. It is defined as 'the percentage of deliveries attended by personnel trained to give the necessary supervision, care, and advice to women during pregnancy, labour, and the postpartum period, to conduct deliveries on their own, and to care for the newborn and the infant', thereby including, beside direct benefits for maternal and child health, transmission of knowledge to mothers (World Bank 2000, notes on Table 2.16). Quite comprehensive data on the proportion of births attended by health staff for 1998 were released by the World Bank (2000). Available data for the years 1990-98 indicate that rapid changes in this variable rarely occur. The data referring to 1998 can be assumed to be representative for the year 1997 or even the period 1995-97. Except for four countries, data for 1998 are available for all nations with NI 1997 in the basic data set[26]. For countries with NI 1992, the most recent information from the period 1988-92 was used[27]. Consequently, data are available for 92 countries in the basic sample.

Health Environments

The **percentage of the population with access to safe water** is defined as 'the share of the population with reasonable access to an adequate amount of safe water (including treated surface water and untreated but uncontaminated water, such as from springs, sanitary wells, and protected boreholes). An adequate amount of safe water is that needed to satisfy metabolic, hygienic, and domestic requirements—usually about 20 litres a person a day' (World Bank 2000, notes on Table 2.14). According to the customary definition, the **percentage of the population with access to sanitation** is 'the share of the population with at least adequate excreta disposal facilities that can effectively prevent human, animal, and insect contact with excreta. Suitable facilities range from simple but protected pit latrines to flush toilets with sewerage. To be effective, all facilities must be correctly constructed and properly maintained' (World Bank 2000, notes on Table 2.14). Data for both indicators are taken from UNICEF's report on the State of the World's Children 1999, containing the latest information available in the period 1990-97 (UNICEF 1999). Therefore, proper matching with NI reference years 1992 and 1997 is not possible[28.] Inclusion of access to safe water or

---

[26] The exceptions are China (1994), Kazakhstan (1995), Mali (1996) and Sri Lanka (1993). The figures in parentheses indicate the years from which data on the percentage of births attended by health staff were taken in the absence of more recent information.

[27] For nine out of 15 countries with NI 1992, data from 1990, 1991 or 1992 are available. For five other countries (Ecuador, Ethiopia, Gabon, Guinea, Trinidad and Tobago) figures for 1988 or 1989 are used. As far as Romania is concerned, information is only available for 1996, indicating the attendance of health staff in 99% of all births. This figure has been incorporated into the data set also, because it is likely that the actual percentage around 1992 was at a comparatively high level.

[28] Using data published by the World Bank (1999) is not a better solution that could help to avoid this problem. Data availability is scanty for given years, and in order to preserve a reasonable number of

access to sanitation as independent variables restricts the number of countries left in the regression to 89 and 87, respectively.

Another variable relating to health environment is the **percentage of the population with access to health care**, which is defined as 'the share of the population that can expect treatment for common diseases and injuries, including essential drugs on the national list, within one hour's walk or travel' (World Bank 2000, notes on Table 2.14). The data for this variable were compiled from the most recent information on access to health care from 1985-93[29] (World Bank 1998). Similar problems with scanty data availability, no possibility of proper matching with NI data, and occasional quick changes over time as for the other two access rates exist. As a result, data for access to health care are only available for 59 countries in the sample.

In contrast to access to health care, 'urbanisation' is used not only as a proxy variable for access to public services in general, but is also meant to capture to some extent the quality of health care, which is thought usually to be better in urban than in rural areas. The operational measure of urbanisation is the **percentage of the (mid-year) population living in urban areas**, which has the advantage of broad data availability. As urbanisation develops slowly and very steadily, no three-year averages were formed, but data are taken for the years 1992 and 1997 from World Bank (2000), and matched with NI reference years.

HIV/AIDS

Data by country on the **prevalence of HIV infection and AIDS in adults** have been regularly published by WHO since June 1998 (WHO 2000b). However, they were not made available by country for the early 1990s, when estimates were given only by region. Therefore, it is not possible to match NI values for 1992 properly with data on HIV/AIDS. Prevalence rates pertaining to the end of 1997 have been employed for countries with NI 1992 and countries with NI 1997 alike. Doubts can be raised about the precision of these broadly available estimates gained from computer simulations of epidemiological models, because the figures are usually based on infection prevalence

---

countries for the analysis, the data have to be compiled from several years, as has been done in UNICEF 1999. However, in some instances, rapid changes in access to safe water or sanitation can be observed over time (World Bank 1999), which casts doubts on the validity of the variables compiled from a range of years.

[29] The 1999 and 2000 editions of the World Development Indicators CD-ROM contain no data about access to health care, and on the 1998 edition, no information later than 1993 can be found.

data collected anonymously at health facilities, whereas community-based surveys in rural areas are still 'sorely lacking' (UNAIDS/WHO 1998, p. 52).

7.3 Estimation Strategy and Results

As already mentioned, the aim of the present chapter is a crude assessment of main NI determinants. Deeper insights about the interrelationship of NI components and their explanatory factors cannot be gained in a regression of the NI as an aggregate. However, some overarching political, economic, and social variables acting simultaneously on several aspects of the nutrition situation can be identified by means of this rough approach. A distinction between basic and underlying determinants is not made during the estimation procedure. Therefore, partial rather than total effects are detected by the model. For example, if GNP per capita is used in combination with institutional variables (e.g. political freedom index or ICRG country risk indices), the total effect of the political indicators on food security and nutrition is probably underestimated, because democracy and good governance are likely to influence economic growth[30]. Similarly, it is not possible to explore how economic resource availability or political freedom or good governance work *through* the provision of public services, if the respective variables are combined in the same regression. Instead of this, the coefficient of the GNP per capita variable will reflect its impact on the NI *besides* acting through factors like education and health care provision. In addition, possible interactions of explanatory variables are not considered – thus neither the role of worsening economic conditions and widespread poverty in triggering off violent conflict, nor the impact of warfare on economic development can be assessed. Furthermore, no panel data set is used, and the applied ordinary least squares (OLS) estimation technique cannot account for country fixed effects. If unobserved country fixed effects are correlated with explanatory variables, parameter estimates will be biased. These limitations of the estimation strategy should be kept in mind in the discussion of the results. The simple approach that is chosen, however, is deemed appropriate for the modest purpose of the investigation.

After an initial model has been presented and different war variables have been scrutinised, a basic model is built. Regression of single NI components on the explanatory variables allows the origins of their size and significance in the basic NI model to be identified. Regional dummy variables are discussed in detail, together

[30] A vast body of literature about the impact of institutions on growth exists. However, integrating respective modelling approaches into the present estimation in order to account more fully for the effect of institutional indicators is far beyond the scope of this study.

with the contribution of supplementary variables. Finally, the results of the estimation process are summarised.

7.3.1 Initial Model and Exploration of War Variables

Table 7.3.1-1 shows an initial model with the independent variables bearing the highest explanatory power and significance. This statement applies to various combinations of the variables suggested in the previous part. The correlation matrix of the NI and its explanatory variables is depicted in Table A 7.3.1-1 in the Appendix. The continuous variables and a war dummy variable are supplemented by significant regional dummies. The independent variables are listed in descending order of their relative impact on the Nutrition Index - indicated by the size of the standardised coefficient - with the regional dummies and the intercept in the bottom of the table (deviating from the sequence of hypotheses in 7.1 and of the variable descriptions in 7.2, the explanatory variables are discussed in this order in the following section). The coefficient of determination amounts to 0.876 and thus indicates a notable goodness of fit. The F test as a measure of overall significance of the model is at the same time a test of significance of R-squared (Gujarati 1995). It is highly significant with a rounded p-value of 0.000 for the model presented in Table 7.3.1-1, and likewise for all other regressions discussed in this chapter[31]. The validity of the F test, and the t tests performed on the parameters of the independent variables for hypothesis testing, hinges on a normal distribution of error terms – which is, however, likely for a sample size of 94. Nevertheless, the Kolmogorov-Smirnov-test of normal distribution was run with the residuals of the initial model. The null hypothesis that the residuals are *not* normally distributed was rejected with a p-value of 0.959 (the Kolmogorov-Smirnov-Z value is 0.51).

Regarding the explanatory variables, it is no surprise that logarithmic GNP per capita is the most powerful determinant in the model, which was already evident from the fact that this variable alone accounts for 74% of NI variation, see Chapter 6.3.1. If the logarithm of GNP per capita is used in isolation – thereby capturing the effects of correlated explanatory factors - an increase by one unit corresponds to a gain in NI of 10.99. Yet, the partial effect of the logarithm of GNP per capita is much smaller: when the additional independent variables shown in Table 7.3.1-1 are introduced, the coefficient declines to 6.52. Among the health and human resource related variables, urbani-

[31] As the significance of the F statistic is therefore not able to differentiate among the models, it is not considered any further in the following discussion.

sation and immunisation of children against DPT[32] proved highly significant, differences of ten percentage points in their shares are associated with an increase in NI values of 1.4 and 1.1, respectively.

Table 7.3.1-1 Initial model with war dummy variable relating to a ten-year reference period

Dependent variable **Nutrition Index (NI) 1997 supplemented by NI 1992**				
Independent variables	**Coefficients**			
	non-standard.	**standard.**	**T-statistics**	**Significance**
logarithm of GNP per capita in constant 1995 PPP$	6.52	0.51	8.53 ***	0.000
urban population (in % of total population)	0.14	0.27	4.79 ***	0.000
one-year olds fully immunised against DPT[1] (in %)	0.11	0.21	4.32 ***	0.000
military expenditure (in % of GNP)	-0.46	-0.15	-3.53 ***	0.001
war dummy[2] (0 for non-war countries)	-2.99	-0.12	-3.12 ***	0.003
political freedom index	0.51	0.09	1.96 *	0.053
Regional dummies				
Middle East and North Africa	4.50	0.14	3.02 ***	0.003
South Asia	-3.97	-0.08	-2.09 **	0.039
Eastern Europe and former Soviet Union	3.82	0.10	2.44 **	0.017
Intercept	17.04		3.37 ***	0.001
Model statistics				
Number of observations	94			
R-squared, adjusted	0.876			
Standard error	3.760			
F statistic	74.101			

Source author's own calculations, based on data described in Chapter 7.2

Notes 1 Diphtheria, Pertussis, Tetanus
2 refers to wars begun, ended or continued in 1983-92 or 1988-97 (matched with NI data)

Another health variable that has been hypothesised to have an impact on the nutrition situation, access to health care, showed highly insignificant in whatever combination

[32] Due to multicollinearity, it was not possible to employ the two immunisation variables (relating to DPT and measles) at the same time without rendering them both insignificant. In fact, the Pearson's correlation coefficient for the two variables is 0.92 (p-value = 0.000), which is much higher than any of the correlations between the independent variables actually retained in the initial model, see the correlation matrix in Table A 7.3.1-1 in the Appendix. It turned out when the immunisation variables were used alternatively in the initial model, that immunisation against DPT had higher explanatory power. One might hypothesise that this arises from the fact that full immunisation against DPT requires three sequential doses of the vaccine and immunisation against measles only one. Therefore, three interactions with the health care provider are necessary to immunise a child against DPT, and only one to protect him or her from measles. This means that a better compliance, associated with access to care and parental knowledge and commitment, is needed for full immunisation against DPT: however, as the difference between the impact of the two immunisation variables is rather small, no firm conclusions can be drawn in this respect.

of possible explanatory variables. Remembering the limited ability of this variable to capture the actual benefits of health care and the weakness of its database (see Chapters 7.1.2 and 7.2.2), this is not too surprising. Moreover, HIV/AIDS had no significant partial effect on the NI, but detrimental effects of the pandemic may have been captured by other variables, e.g., GNP per capita[33]. As regards care factors, both variables intended to be proxies for discrimination against females proved insignificant. Educational indicators also showed a disappointing track record vis-à-vis other variables. Their role is considered in more detail in the context of supplementary variables.

Concerning the political variables, three of them were found to be significant. Out of the two Gastil indices, the civil liberties index showed as insignificant in the model. In contrast, the political freedom index was retained as a significant independent variable. However, the estimation result for this variable is in contradiction to the hypothesis stated in Chapter 7.1.3: a rise of two points in the seven-point Gastil index - indicating less political freedom - is associated with an increase of the NI by 1.02. There may be several explanations for this irritating finding. Firstly, less democratic countries may be more likely to fake their official statistics. Secondly, some states with authoritarian regimes subsidise staple foods, eventually with the intention to reduce incentives for civil unrest. However, the relative impact of the political freedom index is quite small as compared to other determinants, and less robust to changes in the set of explanatory variables and countries analysed (see below).

Furthermore, high levels of military expenditure, that have been characterised as an expression of non-beneficial and potentially harmful use of societal resources in Chapter 7.1.1, tend to reduce NI values by five points if they are raised by ten percentage points. Within the country sample, the proportion of military spending varies considerably, ranging from 0.3% of GNP in Mauritius to over approximately 20% in Angola and to nearly 28% of GNP in Saudi Arabia. It is interesting to see that the military expenditure variable is significant even when employed together either with one of the war dummies or with one of the war duration variables with different reference periods. This is the case despite the significant difference in average military expenditure between war and non-war countries[34]. An explanation might be that the two variables

---

[33] Furthermore, the reliability of prevalence estimates has its limits and proper matching with NI 1992 values was not possible, see Chapter 7.2.2. However, the HIV/AIDS variable remains insignificant if used in a regression of NI 1997 values only.

[34] If countries are classified according to experiencing times of war or lasting peace within the ten-year reference period, the average military spending is 5.7% of GNP in war countries and 3.2% in non-war

are relevant for different parts of the country sample. As further analysis shows, the military spending variable is significant for non-war countries (see Table 7.3.1-2) but not within the subset of war countries encompassing a much smaller number of countries (Table 7.3.1-3 on page 181). Whereas diversion of resources to military purposes is likely to be a concurrent harmful factor during wars, high military spending seems to do some damage even to nations living in longer-lasting peace.

Table 7.3.1-2 Initial model for a subset of countries in peace during the ten-year reference period (1983-92 for NI 1992 and 1988-97 for NI 1997)

Dependent variable **Nutrition Index (NI) 1997 supplemented by NI 1992**				
Independent variables	Coefficients			
	non-standard.	standard.	T-statistics	Significance
logarithm of GNP per capita in constant 1995 PPP$	6.24	0.55	7.04 ***	0.000
urban population (in % of total population)	0.13	0.28	3.73 ***	0.000
one-year olds fully immunised against DPT[1] (in %)	0.10	0.20	3.10 ***	0.003
military expenditure (in % of GNP)	-0.50	-0.17	-2.69 ***	0.009
political freedom index	0.56	0.11	1.79 *	0.079
Regional dummies				
Middle East and North Africa	5.08	0.16	2.50 **	0.015
South Asia	-5.39	-0.14	-2.44 **	0.018
Eastern Europe and former Soviet Union	2.49	0.06	1.09	0.280
Intercept	20.27		3.38 ***	0.001
Model statistics				
Number of observations	70			
R-squared, adjusted	0.823			
Standard error	3.925			
F statistic	41.002			

Source author's own calculations, based on data described in Chapter 7.2

Notes As only non-war countries are included here, the war dummy was zero throughout the sample and, therefore, meaningless. The regional dummy for Eastern Europe and the former Soviet Union became insignificant – only five out of the nine nations concerned are retained in the sample, because the other four have been classified as war countries. (Only significant variables are in bold type.)

1 Diphtheria, Pertussis, Tetanus

countries. The difference proved significant in a two-tailed t-test with a p-value of 0.003. If the military expenditure variable is used together with the other determinants in the initial model except for the war dummy, the size of its coefficient increases to –0.54 and the significance changes to a p-value of 0.000, indicating a partial compensation of the omission of the war variable.

Regarding the two war dummy variables with a five-year and ten-year reference period, respectively, both have been employed alternatively in the estimation process. According to the ten-year war dummy, six countries are classified as war countries in addition to the 18 states identified as war countries within the five-year period. This ten-year war dummy has higher explanatory power than the five year dummy, suggesting that even wars which ended more than five years before the NI reference year seem to cast a long shadow on food security and nutrition. Considerable lagged effects are in accordance with the hypotheses stated in Chapter 7.1.1. Yet the difference between the coefficients and significance levels of the two variables is very small (compare Table 7.3.1-1 and Table A 7.3.1-1 including the five-year war dummy in the Appendix) – *ceteris paribus*, the NI is about three points lower for war than for non-war-countries according to both dummies. Adjusted R-squared values of the models in Tables 7.3.1-1 and A 7.3.1-1 show only a slight difference of 0.004. The next question to be addressed now is whether the variables trying to capture the severity of wars – the two war duration variables corresponding to the war dummies – are useful as independent variables. This is tested first for a subsample of war countries.

Indeed, the duration of wars continuing or having ended in the period under consideration (five or ten years back) has a significant impact on the NI, whereas some other variables, including the regional dummies, become insignificant for the very limited sample of countries under consideration[35]. The resulting model for the 24 countries at war within the ten-year reference period is shown in Table 7.3.1-3 (for the variable referring to the five-year period, see Table A 7.3.1-2 in the Appendix). The coefficient of the war duration variable indicates that ten years of warfare cause a loss of 2.6 points in the NI. R-squared and adjusted R-squared have risen considerably vis-à-vis the basic model depicted in Table 7.3.1-1. However, this cannot be mainly attributed to the introduction of the war duration variable. For the given sample of 24 countries, GNP per capita alone accounts for much more of NI variation (namely, 86%) than for the total sample (74%). If added to the other three variables shown in Table 7.3.1-3, the war duration variable in fact increases the adjusted coefficient of determination only from 93.8% to 95.5%.

[35] It may be noted that military expenditure is among the variables that lost their significance. Even if the war duration variable is omitted, significance of the military spending variable is not regained.

Table 7.3.1-3 Model including war duration variable for a subset of war countries according to the ten-year reference period

Dependent variable				
Nutrition Index (NI) 1997 supplemented by NI 1992				
Independent variables	Coefficients			
	non-standard.	standard.	T-statistics	Significance
logarithm of GNP per capita in constant 1995 PPP$	8.20	0.53	6.04 ***	0.000
urban population (in % of total population)	0.17	0.29	4.02 ***	0.001
one-year olds fully immunised against DPT[1] (in %)	0.12	0.24	3.64 ***	0.002
duration of war[2] in years (0 for non-war countries)	-0.26	-0.14	-2.94 ***	0.008
Intercept	3.59		0.47	0.643
Model statistics				
Number of observations	24			
R-squared, adjusted	0.955			
Standard error	2.805			
F statistic	124.106			

Source author's own calculations, based on data described in Chapter 7.2

Notes 1 Diphtheria, Pertussis, Tetanus
2 refers to wars begun, ended or continued in 1983-92 or 1988-97 (matched with NI data)

The duration of wars in years has been used in the present model without any transformation. Yet it might be supposed that allowing for diminishing marginal effects of war years by using a regressive transformation function would increase the explanatory power of the variable. Therefore, the regression was re-run, first, by replacing the simple linear variable by the logarithm of war duration (for countries with one year of war, this transformation results in a variable value of zero) and, second, by the radix of war years. The assumption of declining marginal impact of war years on the NI was not confirmed. In contrast, adjusted R-squared dropped by 0.7 and 0.3 percentage points with the use of the logarithm and the radix of war duration, respectively, and the significance levels decreased (p-values = 0.040 and 0.018, respectively, as compared to 0.008 previously). Subsequently, a possibly *increasing* marginal effect of war duration on the NI is taken into consideration by employing the square of war years. Although this does not seem likely from a theoretical point of view, R-squared actually rises from 0.955 to 0.958 if the square of war duration is used in the regression, and the p-value decreases to 0.004. Though the difference to the result with the simple linear war duration variable is small, the relatively higher explanatory power of squared war years seems to imply that wars are the more destructive for food security and nutrition the longer they last. However, a closer look at individual countries is necessary to make sure that this result does not simply arise from a difference in the nature of wars in the long and short war duration segments of the sample. Indeed, the countries experiencing thirteen years of war and more have been with no exception torn by cruel civil or guerrilla wars. These are – in ascending order of war duration - Sudan, Cambodia, Peru, Mozambique, Angola, Guatemala, Ethiopia, Indonesia and Eritrea. Nevertheless, the same type of extremely detrimental internal wars is found in some, though not all, countries with four years of war duration and less. Respective examples are the Democratic Republic of Congo, Burundi and Sierra Leone, again listed in ascending order of war duration[36].

Besides different general characteristics of protracted wars, the hypothesised cumulative disinvestment in and destruction of human resources in long-lasting wars, as well as possible lagged effects, may explain this result (see Chapter 7.1.1). The functioning of agricultural input and output markets may become more disturbed the longer they have been affected by the fighting. Psychosocial distress is likely to rise with every year of conflict, affecting caring capacity to an increasing extent. The full impact of neglecting education is felt after a whole generation with poor schooling has grown up. Diseases need time to spread after the health care system of a war-ridden country has broken down. Firm conclusions about the observed small increasing marginal effects of war duration and clarification of the reasons for this would, however, require panel data suitable for a country fixed effects analysis. But even if the present analysis is too sketchy to state increasing marginal effect of war years definitely, there is at least no indication that the detrimental impact of war years diminishes with continued war duration.

7.3.2 Basic Model and Regressions of NI components

The initial model is slightly modified by replacing the war dummy by a war duration variable (with a variable value of zero for non-war countries, thereby assuming no effect for this group), which is justified by the preceding separate investigation of war countries. This model, which is shown in Table 7.3.2-1, is subsequently termed the basic model. Adjusted R-squared has remained virtually unchanged as compared to the

[36] A war duration of one year is assigned to Burundi and Sierra Leone, because for these countries NI 1992 was used since data availability did not allow the calculation of NI 1997. Liberia and Somalia could be added to the list of countries with cruel civil wars of short duration in 1992, but they were not part of the sample due to lack of GNP data.

initial model with the war dummy (compare Table 7.3.1-1)[37]. There has been little change with respect to the coefficients and significance levels of the variables, either (naturally, the coefficient of the war variable does not remain the same when shifting a dummy to a continuous variable). A noteworthy difference to the initial model is the lower regional dummy for Eastern Europe and the successor states of the Soviet Union, which nearly becomes insignificant. This may be explained by the fact that four out of the nine countries in this region are war countries, with a lower mean war duration than the average of all war countries (4 years as opposed to 9.75 years)[38]. Part of the size and significance of the regional dummy in the initial model may arise from the lack of further differentiation of war countries, which changes with the introduction of the war duration variable in the basic model.

Table 7.3.2-1 Basic model including war duration variable

Dependent variable **Nutrition Index (NI) 1997 supplemented by NI 1992**				
Independent variables	**Coefficients** non-standard.	standard.	**T-statistics**	**Significance**
logarithm of GNP per capita in constant 1995 PPP$	6.73	0.53	8.87 ***	0.000
urban population (in % of total population)	0.14	0.26	4.60 ***	0.000
one-year olds fully immunised against DPT[1] (in %)	0.11	0.22	4.49 ***	0.000
military expenditure (in % of GNP)	-0.43	-0.14	-3.23 ***	0.002
duration of war[2] in years (0 for non-war countries)	-0.23	-0.12	-3.07 ***	0.003
political freedom index	0.54	0.09	2.06 **	0.043
Regional dummies				
Middle East and North Africa	3.66	0.11	2.43 **	0.017
South Asia	-4.20	-0.09	-2.21 **	0.030
Eastern Europe and former Soviet Union	2.62	0.07	1.69 *	0.094
Intercept	15.07		3.02 ***	0.003
Model statistics				
Number of observations	94			
R-squared, adjusted	0.876			
Standard error	3.766			
F statistic	73.822			

Source author's own calculations, based on data described in Chapter 7.2

Notes 1 Diphtheria, Pertussis, Tetanus
2 refers to wars begun, ended or continued in 1983-92 or 1988-97 (matched with NI data)

[37] Again, the Kolmogorov-Smirnov-test was performed to test for the normal distribution of residuals, which was confirmed with an almost identical test statistic (0.52) and p-value (0.951, expressing the probability of error of the null hypothesis of no normal distribution) as for the initial model.

[38] The significance of the regional dummy for Eastern Europe and the successor states of the former Soviet Union indeed seems to hinge on the four war countries, as the dummy is insignificant if only non-war countries are considered, compare Table 7.3.1-2.

The meaning of the three significant dummy variables is discussed in more detail below in the context of additional variables and of single indicator models. Before turning to supplementary variables in Section 7.3.3, the impact of the significant variables on the single indicators constituting the NI is explored to check the hypotheses laid out in Chapter 7.1, which are based on the components of the NI. This approach surely has its limits, because the disaggregated models (see Tables 7.3.2-2, 7.3.2-3 and 7.3.2-4) have not been tailored to explain the variation of the single indicators, and will suffer from omitted variable bias to some extent. However, it is worthwhile identifying which relationships between explanatory variables and single indicators are responsible for the significance of a variable in the NI model.

Table 7.3.2-2 Undernourishment regressed on basic model variables

Dependent variable **Undernourished 1995-1997 supplemented by Undernourished 1990-92 (matched with NI data)**				
Independent variables	**Coefficients**			
	non-standard.	**standard.**	**T-statistics**	**Significance**
logarithm of GNP per capita in constant 1995 PPP$	-9.81	-0.51	-5.26 ***	0.000
urban population (in % of total population)	-0.14	-0.17	-1.87 *	0.065
one-year olds fully immunised against DPT[1] (in %)	-0.16	-0.21	-2.60 **	0.011
military expenditure (in % of GNP)	0.72	0.15	2.19 **	0.031
duration of war[2] in years (0 for non-war countries)	0.47	0.16	2.53 **	0.013
political freedom index	-1.23	-0.14	-1.91 *	0.059
Regional dummies				
Middle East and North Africa	-8.54	-0.17	-2.31 **	0.024
South Asia	-3.53	-0.05	-0.75	0.453
Eastern Europe and former Soviet Union	0.70	0.01	0.18	0.855
Intercept	119.45		9.74 ***	0.000
Model statistics				
Number of observations	94			
R-squared, adjusted	0.673			
Standard error	9.252			
F statistic	22.228			

Source author's own calculations, based on data described in Chapter 7.2

Notes only significant variables are in bold type

1 Diphtheria, Pertussis, Tetanus

2 refers to wars begun, ended or continued in 1983-92 or 1988-97 (matched with NI data)

Table 7.3.2-3 Underweight in children regressed on basic model variables

Dependent variable				
Underweight in Children 1993-1998 (latest)				
supplemented by Underweight in children 1987-92 (latest, matched with NI data)				
Independent variables	**Coefficients**			**Signifi-**
	non-standard.	**standard.**	**T-statistics**	**cance**
logarithm of GNP per capita in constant 1995 PPP$	-5.51	-0.34	-4.10 ***	0.000
urban population (in % of total population)	-0.26	-0.38	-4.97 ***	0.000
one-year olds fully immunised against DPT[1] (in %)	-0.08	-0.12	-1.70 *	0.094
military expenditure (in % of GNP)	0.35	0.09	1.49	0.140
duration of war[2] in years (0 for non-war countries)	0.28	0.11	2.11 **	0.038
political freedom index	-0.22	-0.03	-0.48	0.631
Regional dummies				
Middle East and North Africa	-0.75	-0.02	-0.28	0.780
South Asia	18.26	0.30	5.43 ***	0.000
Eastern Europe and former Soviet Union	-6.56	-0.13	-2.39 **	0.019
Intercept	79.64		9.02 ***	0.000
Model statistics				
Number of observations	94			
R-squared, adjusted	0.764			
Standard error	6.661			
F statistic	34.425			

Source author's own calculations, based on data described in Chapter 7.2

Notes only significant variables are in bold type

1 Diphtheria, Pertussis, Tetanus

2 refers to wars begun, ended or continued in 1983-92 or 1988-97 (matched with NI data)

Interestingly, adjusted R-squared of the NI model exceeds not only the weighted sum of R-squared values of the single indicator models (which would amount to 0.711 if aggregated by means of NI component weights), but is higher than any of them. The residuals of the NI component regressions are normally distributed according to the Kolmogorov-Smirnov-test[39] and the Pearson's correlation coefficients for the residuals are low and insignificant (see Table A 7.3.2-1 in the Appendix). This suggests that the hypothesis of independently distributed (that is, normally distributed and uncorrelated) error terms formulated in Chapter 7.1.3 and the equations derived under this assumption are correct. The coefficients of the NI model, including the intercept, are in fact very well predicted by the Formulas (7.6) and (7.7), with deviations from actual values ranging from –0.16% to +0.05% (see Table A 7.3.2-2 in the Appendix for details).

[39] Though it is supposed that these models suffer from omitted variable bias, the null hypothesis of no normal distribution was rejected by the test with p-values of 0.667, 0.839 and 0.339 for the residuals of the PUN, CUW and CM model, respectively.

Table 7.3.2-4 Under-five mortality regressed on basic model variables

Dependent variable				
Under-five mortality 1997				
supplemented by Under-five mortality 1992 (expressed in %, matched with NI data)				
Independent variables	**Coefficients**			
	non-standard.	**standard.**	**T-statistics**	**Significance**
logarithm of GNP per capita in constant 1995 PPP$	-4.95	-0.58	-6.38 ***	0.000
urban population (in % of total population)	-0.02	-0.06	-0.76	0.451
one-year olds fully immunised against DPT[1] (in %)	-0.10	-0.30	-3.97 ***	0.000
military expenditure (in % of GNP)	0.23	0.11	1.72 *	0.090
duration of war[2] in years (0 for non-war countries)	-0.04	-0.03	-0.49	0.625
political freedom index	-0.17	-0.04	-0.64	0.521
Regional dummies				
Middle East and North Africa	-1.71	-0.08	-1.11	0.270
South Asia	-1.38	-0.04	-0.71	0.482
Eastern Europe and former Soviet Union	-2.16	-0.08	-1.36	0.177
Intercept	57.20		11.20 ***	0.000
Model statistics				
Number of observations		94		
R-squared, adjusted		0.712		
Standard error		3.852		
F statistic		26.542		

Source author's own calculations, based on data described in Chapter 7.2
Notes only significant variables are in bold type
1 Diphtheria, Pertussis, Tetanus
2 refers to wars begun, ended or continued in 1983-92 or 1988-97 (matched with NI data)

Comparing the models for the single indicators with the hypotheses stated in Chapter 7.1 reveals a few surprises. First, urbanisation is insignificant in the regression of child mortality, but shows a negative coefficient and significance at the ten per cent level in the model for undernourishment. A relatively high share of urban residents in the population may presuppose a productive agriculture and/or an economy able to demand food on international markets, and possibly facilitate (emergency) food supplies. Therefore, urbanisation can be considered as a proxy for factors that are favourable to reducing undernourishment. Arguing the other way round, that food insecurity accelerates migration to the cities, is not supported by the present empirical findings[40]. Second, the immunisation variable that was meant to reflect care, knowledge and children's health is found to be significant for undernourishment as a dependent variable.

[40] Other factors being equal, a positive relationship between urbanisation and undernourishment would be expected in this case, and the urbanisation variable would then be endogenous in the PUN model.

However, as this variable is also an indication of an effectively functioning health care system, and may be related to political commitment to the provision of public goods, this result does not seem so odd at second glance. The third surprise is that the war duration variable is not significant with relation to child mortality. The plausible assumption that the significance of this variable might arise from direct killings of children in wars (a kind of death that is not linked to nutritional deficiencies) thus turns out to be wrong[41]. The fourth noteworthy result is that the odd positive sign of the political freedom index in the NI model merely results from its effect on undernourishment: a two-point decrease of the index (indicating more freedom) is associated with a 2.46 percentage points rise in the proportion of undernourished. Considering that Smith and Haddad (2000) successfully employed a combined index of political freedom and civil liberties to explain underweight in children, an impact of the political freedom index on child malnutrition seemed likely. However, they used not only a different sample of countries, but also a panel data set and a different estimation strategy: country fixed effects, separating the basic determinants (GDP per capita and the freedom index) from the so-called underlying determinants. Thus, there are several reasons for divergent outcomes.

Regarding the regional dummies, a look at the single indicator models helps to identify possible reasons for their significance. The regional dummy for the Middle East and North Africa gains its significance mainly from the undernourishment component of the NI; controlling for the impact of other explanatory factors, the share of undernourished is 8.54 percentage points lower in this region than in the rest of the sample. This is not astonishing, because dietary energy supply per capita is quite high in these countries with respect to their level of economic development. Favourable income distribution and comparably little incidence and depth of poverty can be assumed to be responsible for this observation, which is discussed in more detail with respect to additional variables inserted into the basic model.

The two other significant regional dummies mainly arise from the child malnutrition component of the NI. The huge positive dummy for South Asian countries, which amounts to 18.26 in the model presented in Table 7.3.2-3, is a common feature of regressions with underweight in children as the dependent variable. In the models

---

[41] A possible reason for this observation could be that the war variable is based on a long time period preceding the reference year of child mortality. Lagged effects on child mortality might be smaller than presumed, or at least smaller than the impact of past wars on undernourishment and child malnutrition.

designed by Klasen (1999) it amounted to a value of 20, and Smith and Haddad stated that 'if all of the underlying variables were raised to their desirable levels (...) in South Asia, malnutrition would remain at 23.8%, but it would be only 0.5% in Sub-Saharan Africa' (Smith and Haddad 2000, p.93). Various explanations have been proposed for the relatively bad performance of South Asia in child anthropometric indicators, ranging from monsoon climate (FAO 1996a), recurrent flooding and overcrowding as well as other factors facilitating disease transmission, to heavy discrimination against women (Ramalingaswami, Jonsson and Rohde 1996). The possibility of measurement error due to small inadequacies in international child growth standards has also been discussed (Klasen 1999). Further factors may be differences in dietary diversity and micronutrient intakes between the regions. Smith and Haddad made an attempt to consider discrimination against females by using the ratio of female to male life expectancy but, nevertheless, they arrived at the finding on malnutrition in South Asia quoted above. It may be concluded that blaming the comparatively low status of women in South Asia for poor child nutrition is either not the whole truth, or that the variable used is not fully adequate. In the present model for the NI as an aggregate, the proxies employed for discrimination against women proved totally insignificant, but one supplementary variable that is related to discrimination (see Chapter 7.3.3) will help to reduce the South Asia dummy in the regression of child malnutrition at least to some extent.

Referring to the dummy variable for Eastern Europe and the former Soviet Union, it is evident from the partial indicator models that its significance results largely from the impact on child malnutrition (the coefficient amounts to –6.56 in the respective regression), with a possible involvement of the child mortality component of the NI (the coefficient is –2.16 and the p-value 0.177, see the model in Table 7.3.2-4). If the ratio of female to male life expectancy is taken as a yardstick, the status of women can be judged quite favourably in these countries (compare data in WHO 1999), with positive consequences for child nutrition and survival. However, there is another factor that might contribute to the significance of the regional dummy variable: these formerly socialist countries have invested heavily in education over a long period of time. The benefits of this investment are probably still reaped with reference to caring capacity, because human knowledge does not erode as quickly as the economic decay that has been taking place since the breakdown of socialism. Together with Angola and the Democratic Republic of Congo, the successor states of the Soviet Union are the sad leaders of the list of countries with negative growth from the beginning to the end of the nineties. In the Russian Federation, Kazakhstan, the Kyrgyz Republic, Azerbaijan and Armenia, a decline of GNP per capita in PPP$ of between 30% and 60% has been recorded during that period[42] (World Bank 2000). This deteriorating trend will surely have consequences for educational achievements and the human capital stock of the population, though with a considerable time lag. Taking up additional variables next sheds some light on the presumptions stated here concerning the regional dummy variables.

7.3.3 The Contribution of Supplementary Variables

The supplementary variables that are discussed in the following section have several characteristics in common:

1. They account for only small increases in adjusted R-squared when added to the bunch of explanatory variables (which is easily understandable given a value of R-squared as high as 0.876 for the basic model).

2. Data for these variables are not available for all countries in the data set, so restrictions in sample size are unavoidable.

3. The significance of the additional variables is less robust than that of the main determinants presented in Tables 7.3.1-1 and 7.3.2-1, because the supplementary variables become insignificant in certain combinations (the 'births attended by health staff' variable is an exception in this respect, see below).

Indicators of inequality and poverty are not dealt with extensively in the multivariate regression due to limited data availability. Furthermore, their relationship to the NI has been considered in detail in Chapter 6.2 by means of correlation analyses. However, briefly considering indicators of poverty and distribution might help to explain the significance of the positive dummy for the Middle East and North Africa, which arises from the undernourishment component of the NI.

---

[42] These figures are based on the author's own calculations with average GNP per capita values of 1990-92 and 1995-97. The countries listed above are all in the data set. However, there are many more successor states of the Soviet Union without NI values that experienced dramatic decreases in GNP per capita during the nineties, e.g. Tajikistan, Georgia, Turkmenistan, Ukraine, Latvia, Lithuania and Belarus (for Uzbekistan, which is part of the sample, GNP data for the respective periods are incomplete). In contrast, the Eastern European countries in the data set, Croatia and Romania, have been struck by economic stagnation, but not by negative growth.

Figure 7.3.3-1 Box-plot of ranking differences between GNP per capita and the percentage of the population below the poverty line at 1$ a day (for the Middle East & North Africa and other regions)

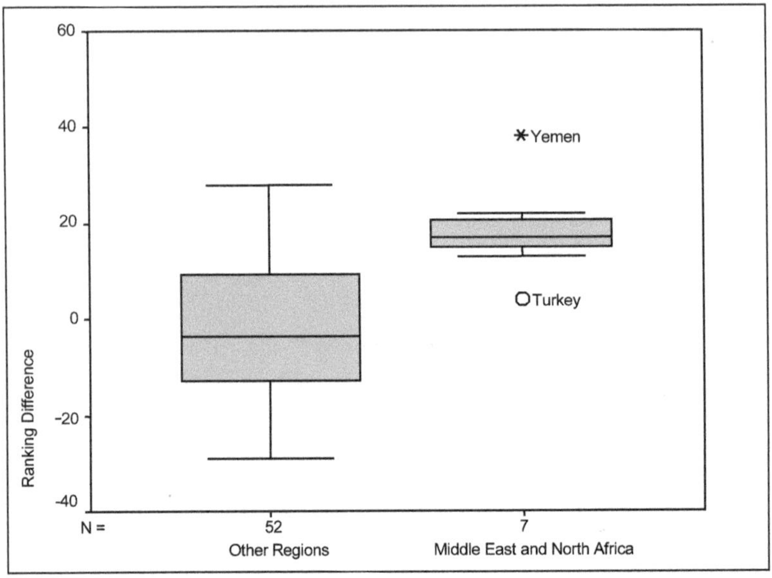

Source author's own presentation based on data described in Chapter 6.2.2

Notes The box covers the middle 50% of the data values between the upper and lower quartiles, the central line is the median. 'Whiskers' extend to the highest and lowest values, unless these are outliers (at a distance of more than 1.5 times the interquartile range from the margin of the box) or extreme values (at a distance of more than 3 times the interquartile range from the margin of the box). Outliers are represented separately by a 'O'-symbol, and extreme values by a star.

Eleven countries in this region are included in the data set, and for seven of them a Gini coefficient is available. Compared to the remaining 64 countries with distributional data in other regions, the seven Middle Eastern and North African countries have a significantly lower average Gini coefficient (37.3 vis-à-vis 44.5 for the average of the other regions), indicating relatively less inequality. At the same time, the incidence of poverty in these countries is very low with reference to their GNP per capita. This is illustrated in Figure 7.3.3-1 for the ranking differences between GNP per capita and the percentage of the population below the poverty line at 1$ a day. Positive values indicate that poverty is less than expected according to GNP per capita (for a more detailed explanation of ranking differences, see Chapter 6.2.2). Similar graphs can be drawn for ranking differences between GNP per capita and the incidence of poverty at 2$ a day and the poverty gap measures.

If the significance of the Middle East and North Africa dummy can actually be traced back to relatively little inequality and poverty in this region, it should become insignificant if the basic regression is run with a distributional measure. However, both measures of distribution described in Chapter 7.2.2 – R20 and the Gini coefficient - miss significance at the 10% level by a small margin, whether added to the basic model or used solely with GNP per capita as a second explanatory variable[43]. Yet, all four international poverty measures (see Chapters 7.2.2 and 6.2.2 for a detailed description of these indicators) prove significant if used together with the GNP per capita variable only. But in the presence of the full range of explanatory variables of the basic model, solely the poverty gap at 1$ a day, indicating the severity of poverty for the lower poverty line, is found to contribute significantly to NI variation. This indicator varies between 0.03% and 39.50%, and *ceteris paribus*, a decrease of ten percentage points raises the NI by 1.1 points. The model with this additional variable is shown in Table 7.3.3-1 on the following page.

The Eastern Europe and former Soviet Union dummy and the military expenditure variable are insignificant for the subset of countries with poverty data, even if no poverty variable is added to the model, and for this reason they are left out. Unfortunately, the Middle East and North Africa dummy is also insignificant for this subsample of 52 countries. A tendency of increasing p-values for this regional dummy can be observed (either being significant at first, as in the country subset with distributional information, or being insignificant from the start, as in the subset with available poverty data), if indicators of poverty or distribution are introduced additionally. However, firm conclusions on the initial hypothesis about the role of poverty and inequality for the significance of the Middle East and North Africa dummy are not in place, because the magnitude of the observed effect is small.

---

[43] Explanations might be found in the non-parametric nature of the Gini coefficient and possible inadequacies in functional form, which, however, do not affect rank correlation analyses as carried out in Chapter 6.3.

Table 7.3.3-1 Model with poverty gap at 1$ a day variable

Dependent variable				
Nutrition Index (NI) 1997 supplemented by NI 1992				
Independent variables	**Coefficients**			
	non-standard.	**standard.**	**T-statistics**	**Significance**
logarithm of GNP per capita in constant 1995 PPP$	6.76	0.53	6.49 ***	0.000
urban population (in % of total population)	0.14	0.25	3.52 ***	0.001
one-year olds fully immunised against DPT[1] (in %)	0.11	0.21	3.37 ***	0.002
duration of war[2] in years (0 for non-war countries)	-0.32	-0.15	-3.22 ***	0.002
political freedom index	0.95	0.15	2.90 ***	0.006
poverty gap at 1$ a day (in %)	-0.11	-0.12	-1.97 *	0.055
Regional dummies				
South Asia	-5.38	-0.15	-2.69 ***	0.010
Intercept	14.53		1.94 *	0.059
Model statistics				
Number of observations	52			
R-squared, adjusted	0.894			
Standard error	3.546			
F statistic	62.603			

Source author's own calculations based on data described in Chapter 7.2

Notes For the limited subset of 52 countries with poverty data, GNP per capita accounted for 78% of NI variation (if used as a single explanatory variable). If the same variables as above, except for the poverty gap indicator, are applied to the present data subset, adjusted R-squared amounts to 0.887, which means that the coefficient of determination increases by only 0.007 when the poverty gap at 1$ a day is added as an independent variable. Military expenditure and the regional dummy variables apart from the South Asia dummy were insignificant for the countries in this subsample (even in the absence of a poverty variable).

1 Diphtheria, Pertussis, Tetanus

2 refers to wars begun, ended or continued in 1983-92 or 1988-97 (matched with NI data)

Another supplementary variable proposed in Chapter 7.1.1 refers specifically to the agricultural sector: the domestic production of cereals, roots and tubers per capita. For reasons which have been outlined in the description of hypotheses, this variable is expected to be insignificant for the total sample of countries. In fact, when added to the basic model for the full data set, the coefficient is almost zero (0.0001) and the p-value is 0.960 (whereas the significance levels of the other explanatory variables remain unchanged, except for the Eastern Europe and former Soviet Union dummy, the p-value of which rises to 0.125). The picture changes when only countries with a GNP per capita of less than 3000 PPP$ are left in the analysis. For this lower income segment in the country data set, the cereals, roots and tubers production variable is highly significant, though the coefficient is low – a rise in staple food production by 100 kg per capita per year would augment the NI by only two points[44]. The estimation result is shown in Table 7.3.3-2, omitting the military expenditure and the three regional dummies which are insignificant due to the restriction of the sample to a subset of countries. The urbanisation variable outperforms the GNP per capita variable with respect to the size of the standardised coefficient, for which no self-evident explanation may be given.

Table 7.3.3-2 Model with cereals, roots and tubers production variable for a subset of lower income countries (GNP per capita < 3000 PPP$)

Dependent variable				
Nutrition Index (NI) 1997 supplemented by NI 1992				
Independent variables	**Coefficients**			
	non-standard.	**standard.**	**T-statistics**	**Significance**
urban population (in % of total population)	0.26	0.42	6.08 ***	0.000
logarithm of GNP per capita in constant 1995 PPP$	6.85	0.40	5.25 ***	0.000
one-year olds fully immunised against DPT[1] (in %)	0.09	0.26	3.48 ***	0.001
duration of war[2] in years (0 for non-war countries)	-0.29	-0.21	-3.30 ***	0.002
cereals, roots & tubers production (in kg/capita/year)	0.02	0.20	3.05 ***	0.004
political freedom index	0.86	0.18	2.73 ***	0.009
Intercept	5.22		0.65	0.521
Model statistics				
Number of observations	55			
R-squared, adjusted	0.805			
Standard error	3.756			
F statistic	38.136			

Source author's own calculations based on data described in Chapter 7.2

Notes 1 Diphtheria, Pertussis, Tetanus
2 refers to wars begun, ended or continued in 1983-92 or 1988-97 (matched with NI data)

The significance of the cereals, roots and tubers production variable still has to be interpreted with caution. Errors occurring in the specification of staple food production will inevitably translate into the dietary energy supply figures and undernourishment estimates. For poor countries, the impact of measurement error in cereals, roots and tubers production figures on dietary energy supply data might be large, because these staples constitute a large share of dietary energy intakes. Furthermore, statistical information from low income countries is, in general, relatively less reliable than from more advanced countries. Thus, the significance of this variable of agricultural pro-

[44] If the cut-off for inclusion of countries is set at 2000 PPP$, this yields similar results which are not shown here.

duction may be at least to some extent caused by correlated measurement errors of the explanatory variable and the undernourishment component of the NI.

Table 7.3.3-3 Basic model supplemented by literacy rate

	Coefficients			
Dependent variable				
Nutrition Index (NI) 1997 supplemented by NI 1992				
Independent variables	**non-standard.**	**standard.**	**T-statistics**	**Significance**
logarithm of GNP per capita in constant 1995 PPP$	6.00	0.47	7.49 ***	0.000
urban population (in % of total population)	0.12	0.23	4.08 ***	0.000
one-year olds fully immunised against DPT[1] (in %)	0.10	0.19	3.89 ***	0.000
military expenditure (in % of GNP)	-0.40	-0.13	-3.05 ***	0.003
duration of war[2] in years (0 for non-war countries)	-0.23	-0.12	-3.09 ***	0.003
political freedom index	0.52	0.09	2.00 **	0.049
Regional dummies				
Middle East and North Africa	4.60	0.14	3.02 ***	0.003
South Asia	-3.53	-0.07	-1.88 *	0.064
Eastern Europe and former Soviet Union	1.76	0.04	1.08	0.284
Added variables				
total literacy rate (in %)	0.06	0.14	2.43 **	0.017
Intercept	18.17		3.60 ***	0.001
Model statistics				
Number of observations	93			
R-squared, adjusted	0.882			
Standard error	3.679			
F statistic	70.075			

Source author's own calculations based on data described in Chapter 7.2

Notes Only significant variables are in bold type.

1 Diphtheria, Pertussis, Tetanus

2 refers to wars begun, ended or continued in 1983-92 or 1988-97 (matched with NI data)

With reference to education, the variables proposed in Section 7.2.2 have been tested: female secondary education (gross enrolment rate), female literacy rate, and total literacy rate. Yet, female secondary education proved insignificant, and female literacy could only gain significance if either the urbanisation variable or the South Asia dummy were omitted. In the presence of the urbanisation variable, female literacy showed to be only weakly significant (at the ten per cent level). This suggests that part of the explanatory power of urbanisation stems from a relation to (female) education. Living in cities and towns can be thought to provide more educational incentives and opportunities, maybe specifically to women. The total literacy rate, however, can be employed in addition to the urbanisation variable whilst retaining its significance, see Table 7.3.3-3 on the previous page.

However, in comparison to the coefficients of urbanisation and immunisation of children, literacy makes only a minor contribution: a gain of ten percentage points corresponds to an increase in the NI of 0.6[45]. The significance of the literacy variable can be traced back to its impact on child malnutrition and mortality, whereas it is insignificant with undernourishment as the dependent variable. Interestingly, the dummy for Eastern Europe and the former Soviet Union becomes insignificant when the literacy variable is added, which supports the explanation suggested above that the significance of this dummy in the basic model is due to favourable achievements in education in this region.

Since only one country had to be dropped from the sample due to lack of data when the literacy variable was introduced, it might be tempting to use this variable as one of the main determinants in the basic model. However, the variable is not as robust as would be desirable, which becomes evident, when 'births attended by health staff (in %)' is employed additionally. In this combination, literacy becomes insignificant (see Table 7.3.3-4 on the following page). As in the case of literacy, the significance of this new variable results from its impact on child malnutrition and mortality only. The South Asia dummy loses its significance for the NI model, too, if the percentage of births attended by health staff is used[46]. But this dummy is still significant if underweight in children is the dependent variable. However, with the inclusion of 'births attended by health staff', it is reduced from a value of 18.16 to 14.71 in the underweight model[47]. As the proportion of births attended by health staff represents public commitment to the well-being of mothers and children, knowledge of mothers and ability to act in their best interest, it may hinge on literacy and capture it to some ex-

---

[45] A noteworthy change taking place with the introduction of the literacy rate as an independent variable is the decline of the parameter of the logarithm of GNP per capita from 6.73 to 6.00, which indicates that this variable captures to some extent effects of literacy in the basic model. The coefficients of the initial two health and human resource related variables are slightly reduced when literacy is added to the set of explanatory variables.

[46] Furthermore, the regional dummy for Eastern Europe and the former Soviet Union is insignificant in the model with the 'births attended by health staff' variable (with and without literacy as an additional variable). However, this seems to be an effect of the reduction in the number of countries in the analysis if this variable is used. If the basic variables only are employed for the 91 countries with data on births attended by health staff (dropping three countries from the regression), the Eastern Europe and former Soviet Union dummy is already insignificant. In contrast, the literacy variable is also significant for this limited subset of countries.

[47] Both figures given are for countries in the sample with data on births attended by health staff only.

tent. High proportions of births attended by trained staff are furthermore likely to be associated with a relatively good status of women and the absence of heavy discrimination, as has been hypothesised in Chapter 7.1.2. This may further explain the ability of the variable to render the South Asia dummy insignificant.

Table 7.3.3-4 Basic Model supplemented by the variables 'literacy rate' and 'births attended by health staff'

Dependent variable				
Nutrition Index (NI) 1997 supplemented by NI 1992				
Independent variables	Coefficients			
	non-standard.	standard.	T-statistics	Significance
logarithm of GNP per capita in constant 1995 PPP$	5.54	0.43	6.66 ***	0.000
urban population (in % of total population)	0.12	0.23	3.96 ***	0.000
one-year olds fully immunised against DPT[1] (in %)	0.09	0.18	3.39 ***	0.001
military expenditure (in % of GNP)	-0.46	-0.15	-3.39 ***	0.001
duration of war[2] in years (0 for non-war countries)	-0.21	-0.11	-2.70 ***	0.008
political freedom index	0.68	0.12	2.55 **	0.013
Regional dummies				
Middle East and North Africa	4.32	0.13	2.83 ***	0.006
South Asia	-2.45	-0.05	-1.24	0.218
Eastern Europe and former Soviet Union	0.69	0.02	0.39	0.700
Added variables				
births attended by health staff (in %)	0.05	0.14	1.83 *	0.072
total literacy rate (in %)	0.03	0.07	1.14	0.258
Intercept	20.81		4.04 ***	0.000
Model statistics				
Number of observations	90			
R-squared, adjusted	0.884			
Standard error	3.642			
F statistic	62.636			

Source author's own calculations based on data described in Chapter 7.2

Notes Only significant variables are in bold type.

1 Diphtheria, Pertussis, Tetanus

2 refers to wars begun, ended or continued in 1983-92 or 1988-97 (matched with NI data)

Other possible supplementary health variables are access to sanitation and to safe water. Though the sanitation variable is able to make a significant contribution and increase adjusted R-squared if added to the basic model, it is not very robust. Significance is lost if the literacy variable is introduced additionally (which is also insignificant in this combination), and the same happens in the presence of the 'births attended by health staff' variable, which, however, retains high significance. The p-value of the safe water variable declines when it is used in the absence of the urbanisation variable, but it still misses significance at the ten percent level. If used together with the sanitation variable, it is pushed far from significance. Due to their lack of robustness, these two variables are not considered any further. Their weakness may partly stem from the lack of matching with NI reference years and the scattering of data over several years, see Chapter 7.2.2 for details.

Table 7.3.3-5 Basic model for a subset of countries with ICRG index data

Dependent variable				
Nutrition Index (NI) 1997 supplemented by NI 1992				
Independent variables	Coefficients			
	non-standard.	standard.	T-statistics	Significance
logarithm of GNP per capita in constant 1995 PPP$	6.55	0.53	7.37 ***	0.000
urban population (in % of total population)	0.13	0.24	3.65 ***	0.001
one-year olds fully immunised against DPT[1] (in %)	0.13	0.25	4.44 ***	0.000
military expenditure (in % of GNP)	-0.40	-0.14	-2.86 ***	0.006
duration of war[2] in years (0 for non-war countries)	-0.15	-0.07	-1.68 *	0.098
political freedom index	0.65	0.11	2.18 **	0.033
Regional dummies				
Middle East and North Africa	3.49	0.12	2.24 **	0.029
South Asia	-5.57	-0.13	-2.58 **	0.012
Eastern Europe and former Soviet Union	3.03	0.05	1.06	0.292
Intercept	15.07		2.63 **	0.011
Model statistics				
Number of observations	75			
R-squared, adjusted	0.871			
Standard error	3.821			
F statistic	56.744			

Source author's own calculations based on data described in Chapter 7.2

Notes Only significant variables are in bold type.

1 Diphtheria, Pertussis, Tetanus

2 refers to wars begun, ended or continued in 1983-92 or 1988-97 (matched with NI data)

Another additional variable to be tested is the ICRG index, which has been chosen to capture various elements of good governance that are hypothesised to improve the nutrition situation. The theoretical maximum value of the index amounts to 50, with higher values indicating better governance. Unfortunately, data are only available for 75 out of the total sample of 94 countries. For a proper assessment of the relative contribution of the ICRG index, the basic model has to be recalculated for this limited sample of 75 countries for comparison, see Table 7.3.3-5. The Eastern Europe and former Soviet Union dummy becomes insignificant and is omitted in the following estimations, and the war duration variable has lost part of its significance and explanatory power (for seven out of 24 war countries in the initial sample, no ICRG data are available).

In the presence of the ICRG index, the war duration variable loses its significance (p-value = 0.128) but the ICRG variable also misses significance by a small margin (p-value = 0.134). In fact, the mean ICRG index value for non-war countries is significantly higher than for war countries[48] – not surprisingly, bad governance and warfare are not independent of each other. Furthermore, when the ICRG index is added to the model, the political sphere seems to be represented by too many variables that are related to each other. If the ICRG variable only is employed together with GNP per capita, it is significant with a p-value of 0.055 and the coefficient amounts to 0.20. Significance at the ten per cent level is also retained if only the social indicators immunisation, urbanisation and the South Asia and Middle East & North Africa dummies are used additionally in the regression.

Table 7.3.3-6 Model with GNP per capita and three-component ICRG index as explanatory variables

Dependent variable: Nutrition Index (NI) 1997 supplemented by NI 1992				
Independent variables	**Coefficients**			
	non-standard.	**standard.**	**T-statistics**	**Significance**
logarithm of GNP per capita in constant 1995 PPP$	9.65	0.79	12.14 ***	0.000
three-component ICRG index[1]	0.36	0.16	2.54 **	0.013
Intercept	0.43		0.79	0.937
Model statistics				
Number of observations	75			
R-squared, adjusted	0.768			
Standard error	5.132			
F statistic	123.505			

Source author's own calculations based on data described in Chapter 7.2

Notes For the present sample of 75 countries, a regression on the logarithm of GNP per capita alone results in an adjusted R-squared of 0.751.

1 comprises indices relating to 'rule of law', 'expropriation risk', 'repudiation of contracts by the government'

[48] The mean ICRG index value for non-war countries is 32.1, and 28.4 for war countries. For equal variances (which are confirmed by the Levene test with a p-value of 0.006), the difference has a two-tailed significance at 4.9%.

Moreover, not all components of the ICRG index may actually have an impact on food security and nutrition as measured by the NI. For this reason, the five single risk indices constituting the ICRG index are checked separately in regressions with GNP per capita as the sole additional variable. The indices for quality of the bureaucracy and corruption are highly insignificant (with p-values of 0.846 and 0.671, respectively) whereas the other three components of the ICRG index reach significance at the five per cent level. The three significant subindices – reflecting rule of law, expropriation risk, and repudiation of contracts by the government - are combined in a tripartite ICRG index[49]. This three-component ICRG index performs better than the full ICRG index when used together with GNP per capita (see the model with the two explanatory variables in Table 7.3.3-6).

Table 7.3.3-7 Model with three-component ICRG index as a governance indicator

Dependent variable Nutrition Index (NI) 1997 supplemented by NI 1992				
Independent variables	**Coefficients** non-standard.	standard.	**T-statistics**	**Significance**
logarithm of GNP per capita in constant 1995 PPP$	6.35	0.52	6.97 ***	0.000
urban population (in % of total population)	0.13	0.25	3.76 ***	0.000
one-year olds fully immunised against DPT[1] (in %)	0.11	0.22	3.70 ***	0.000
military expenditure (in % of GNP)	-0.41	-0.14	-2.98 ***	0.004
political freedom index	0.59	0.10	2.02 **	0.047
three-component ICRG index[2]	0.20	0.09	1.68 *	0.097
Regional dummies				
South Asia	-6.07	-0.13	-2.78 ***	0.007
Middle East and North Africa	2.97	0.10	1.91 *	0.060
Intercept	13.56		2.40 **	0.019
Model statistics				
Number of observations	75			
R-squared, adjusted	0.871			
Standard error	3.824			
F statistic	63.575			

Source author's own calculations based on data described in Chapter 7.2

Notes 1 Diphtheria, Pertussis, Tetanus

2 comprises indices relating to 'rule of law', 'expropriation risk', 'repudiation of contracts by the government'

---

[49] The three-component ICRG index is calculated in the same manner as the ICRG index comprising all five elements of governance: the index reflecting 'rule of law' is first transformed from its 0-6 scale to a 0-10 scale, and then summed up with the other two variables under consideration. It should be noted that the maximum value of the tripartite ICRG index is only 30, because only three subindices are included.

However, the tripartite ICRG index still cannot retain significance in the presence of the war duration variable (which also turns insignificant if used together with the index) or the war dummy variable in the basic model. If the war duration variable is omitted, the reduced ICRG index reaches significance at the ten per cent level (the resulting model is shown in Table 7.3.3-7 on the previous page). Without the Middle East and North Africa dummy, the coefficient of the reduced ICRG index increases to 0.25 and the p-value reaches 0.040, but adjusted R-squared declines to 0.866. The mean value of the three-component ICRG index is actually significantly higher for the Middle East and North Africa than the respective value for the other regions (25.2 versus 21.0, the p-value of the t-test is 0.008). Better governance indicators as compared to the rest of the sample may partly explain the positive regional dummy variable for the Middle East and North Africa.

As has been pointed out in the introduction to this chapter, the full impact of good governance on food security and nutrition cannot be assessed by such a simple estimation strategy, because governance is likely to exert its impact on the NI through other social and economic variables. Moreover, bad governance seems to be associated with belligerent regimes, and may even be the common root cause of hunger and violent conflict. Philippe Guiton has stated that 'Fundamental grievances that motivate violent collective action arise from the same political processes that generate food shortages' (Guiton 2001, p.1). Notably, the significant ICRG subindices all represent elements of legal and economic insecurity, which are at the same time frequent concurrent factors of warfare. In contrast, the insignificant components of the ICRG index relate to weak, inefficient bureaucracies and corruption, which can be considered as calculable evils. Since the three-component ICRG index draws its significance solely from its impact on the undernourishment component of the NI, it can be hypothesised that the climate of insecurity, which this index represents, hampers economic incentives and, therefore, food production, food marketing, and finally food supply.

7. 4 Summary of Results

The main independent variables employed in the basic model are able to explain 88% of NI variation, which constitutes a notable increase vis-à-vis the coefficient of determination of a univariate regression on the logarithm of GNP per capita (that amounts to 74%). The same variables account for no more than 66%, 76% and 71% of variation in the NI components undernourishment, underweight in children and under-five mortality, respectively. As the residuals of the single indicator models are normally distributed and only weakly and insignificantly correlated, the increase in the coefficient of determination of the NI model can be explained by a decline in the variance of independently distributed error terms due to aggregation. Apart from the influence of omitted explanatory variables, the residuals are likely to comprise random measurement errors of the dependent variables.

The logarithm of GNP per capita as an indicator of economic resource availability is the dominating variable in the regression of the NI (its parameter value is 6.73 in the basic model). Regarding other economic variables, common indicators of poverty and distribution do not contribute significantly to NI variation if added to the bunch of political and social variables in the basic model, because their effects seem to be already captured by the main determinants. An exception is the poverty gap at 1$ a day (with a coefficient of –0.11), which is, however, only available for a limited sample of countries, like other indicators of poverty and distribution.

As the sole variable relating specifically to the agricultural sector, domestic staple food production has been employed and proved significant for countries with a GNP per capita below 3000 PPP$; the parameter amounts to 0.02. The result is in line with the expectation that cereals, roots and tubers production would only be significant for a subset of lower income countries. Yet this has to be interpreted with caution, because measurement error in specifying staple food production in these countries may influence NI values and the estimation result via the undernourishment component.

Among the variables relating to health, care and education, urbanisation turns out to have high explanatory power. The coefficient indicates that an increase of ten percentage points in urbanisation raises the NI by 1.4 points. Seemingly, this variable captures to some extent educational opportunities and achievements, and possibly access to safe water, information, infrastructure and quality of health care provision, also. These indicators are in general more favourable in urban than in rural areas. For this reason, urbanisation is a proxy for knowledge and health environments. Its function as a 'multi-purpose' variable is at the same time a strength and a weakness: the advantage is the robustness and high significance of the variable, the disadvantage is the uncertainty about what urbanisation actually represents in the context of food security and nutrition. Other health variables – physical access to health care and HIV/AIDS – proved insignificant. Health variables like immunisation of children and births attended by health staff, which incorporate actual utilisation of services, are more useful in explaining NI variation than mere indicators of access to facilities (like access to health care, safe water or sanitation). Utilisation of health services requires not only physical access, but also the willingness to use them. Compliance implies some knowledge and understanding of the benefits, thereby reflecting caring capacity to a certain extent. Concerning other aspects of care, none of the variables meant to be proxies for discrimination against females was significant. But the supplementary variable 'births attended by health staff' is likely to reflect, among other things, how much special care and assistance is offered to and called on by mothers, thereby referring to women's (and children's) status in society.

Several political variables have been tested in the model. Military expenditure (expressed as a percentage of GNP), which also has, of course, an economic meaning, proved significant. However, this was only the case for non-war countries if the sample was split up. The variable became insignificant for some other subsets of the total data set, which provokes doubts about its reliability as an explanatory factor. Regarding the war variables, the two war dummies and the war duration variables proved significant. Other explanatory factors being equal, the NI is about three points lower for war countries than for non-war countries. The variables including wars in the longer time period of ten years (the NI reference year itself and the nine preceding years) had slightly more explanatory power, suggesting considerable lagged effects. Regarding the war duration variables, ten years of war reduce the NI by more than two points, and no declining marginal effects of war years could be found. Of the Gastil indices relating to democracy, only the political freedom index was significant. with a coefficient of 0.54, showing the unexpected result that *ceteris paribus*, higher NI values are associated with less democracy. The ICRG index employed as a proxy for good governance, which could only be used for a subset of countries due to limited data availability, was merely significant in the absence of war variables. Interestingly, the significance of the ICRG index in the model could be traced back to the components associated with a climate of legal and economic insecurity ('rule of law', 'expropriation risk', 'repudiation of contracts by the government'), but not to the two subindices indicating weak and corrupt bureaucracies. Yet, as already mentioned at the beginning of the chapter, the estimation strategy does not allow assessment of the total effects of democracy, good governance or warfare, which may as well work through other variables in the model (e.g., GNP per capita and indicators of public service provision).

As regards the regional dummy variables, they can at least be partly explained by means of additional explanatory variables, which weaken or extinguish their significance. The Middle East and North Africa is a region with relatively high dietary energy supply per capita and consequently low percentages of undernourishment.

Though the statistical evidence is not unequivocal in this respect, low incidence of poverty in the region relative to GNP per capita, and comparatively good governance as measured by the ICRG index, can be assumed to be responsible for the significant positive regional dummy (the parameter is 3.66 in the basic NI model). The reason for the negative dummy variable of South Asia (-4.20) is totally different: it is obviously an outflow of the region's poor performance in child nutrition. The reasons for the 'South Asian enigma' of extremely high malnutrition prevalence accompanied by quite favourable economic and social indicators are not yet fully clarified, but lack of assistance to mothers and small children, and discrimination against females in general, could explain the puzzle to some extent (Ramalingaswami et al. 1996). In Eastern Europe and the former Soviet Union, the considerable human capital stock with reference to education seems to benefit child nutrition and health even in the face of economic decline, leading to a positive coefficient of 2.62 for the regional dummy variable in the basic model.

8 Discussion

The desirable properties of international indicators and indices which have been in operation prior to the design of the Nutrition Index have been highlighted. Having considered respective requirements throughout the construction and calculation of the NI and having analysed redundancy and sensitivity, an account of the characteristics of the NI is given against this benchmark, followed by a discussion of this new measure in the light of the vivid debate on international indices.

8.1 Evaluation of the NI against the Requirements for International Indicators

The question if the NI can be **policy-relevant** cannot be answered with reference to actual experience, since it has found no broad application to date. The partial indicators of the NI, however, have proved relevant in the decision-making process. Compared to its components, the NI shows the advantage of greater sensitivity to absolute poverty (at least with reference to international data), which could be an important characteristic for targeting projects and programmes[1]. The requirement that international indicators and indices should be **adequate** was considered in the choice of partial indicators for the dimensions of the nutrition situation. As the NI focuses on the measurement of food insecurity and undernutrition, it is surely not adequate for countries where these problems have been almost completely overcome and overnutrition is by far the most prevalent type of malnutrition (in the words of Ryten (2000), the NI does not 'answer the right questions' for these countries). Apart from very limited data availability for industrial countries, this provided a rationale for restricting the calculation of NI values to developing countries and a few countries in transition. Some fast-developing nations and countries in transition where large-scale over- and undernutrition coexist, and for which NI values have been computed, may be regarded as borderline cases. Considering further requirements for a 'high profile single number' stated by Ryten (2000), it is yet to be shown if the NI is **communicable** to politicians and the general public. Though the NI is based on a rather complex concept of food insecurity and undernutrition, its simple construction might be favourable in this respect. In general, single numbers like the NI can be communicated more easily to a

---

[1] In fact, the NI, which has been calculated in subnational disaggregation for India and China, might contribute to recent efforts of poverty mapping at the regional level, see the information posted at http://www.povertymap.net/index.htm for more information on the geographic analysis of poverty and human welfare.

broad audience than a bunch of various indicators. Provided that the underlying data for the percentage of undernourished, underweight in children and child mortality are available, the NI is **replicable** with simple means – there is no need to employ electronic calculating devices[2]. The claim that an international indicator should be **backed by theory** is not easy to fulfil for composite indices like the NI, because there is no unambiguous way to convert available empirical variables into a common measurement unit expressing the abstract variable that is the ultimate criterion of interest[3]. At least, the NI is based on a conceptual framework that is scientifically founded and generally accepted (see Chapter 3.1). Assessing whether the NI is **credible**, which is deemed a desirable property by Ryten (2000), cannot be the task of the author of this study. Credibility might be discussed for individual countries with respect to the underlying data, an undertaking that is not pursued here.

Regarding the technical requirements, **non-redundancy** of the NI vis-à-vis its components and the main development indicators is not fully achieved according to the rank correlations computed in Chapters 6.1.1 and 6.3. Given the unavoidable multicollinearity of international social and economic indicators, this is not surprising. Yet, the NI could be shown to be much less redundant or even non-redundant for subsamples of the total data set. It should be noted that similar levels of redundancy were found for the HDI, GDP per capita and the partial indicators of the Human Development Index for the entire group of countries under review, as were comparably low rank correlation coefficients for the subsamples (McGillivray and White 1993, Noorbaksh 1998). A certain redundancy has in no way invalidated the approach of UNDP to assess the level of human development against the yardstick of economic development. Likewise, the NI has been shown in this study to contain information on food security and nutrition that is not adequately taken into account by GNP per capita, the HDI or any of its partial indicators. An analysis of respective outliers can yield valuable insights, which has been illustrated by reference to selected country cases.

Concerning the **validity** of the NI, it mainly depends on the validity of its components (and on the existence of a conceptual base for aggregation, see above). Whereas it is generally accepted that methods to capture child mortality and anthro-

[2] The Hunger Index, in contrast, involves taking the logarithm of dietary energy supply, which cannot be done that easily without a computer, electronic calculator, or at least a slide rule.

[3] Indices of this type are frequently accused of counting up 'apples and pears' (and, possibly, oranges), but this may still be preferable to talking about apples or pears or oranges only, if fruit is the matter of discussion - i.e. to the exclusive use of a one-sided single indicator.

pometric status are actually measuring what they intend to quantify in both a conceptual and statistical sense[4] (see Ramalingaswami et al. 1996 and WHO 1997 on anthropometry), the validity of FAO's undernourishment indicator has been seriously questioned. Certainly, an assessment of food security relying solely on dietary energy supply relative to requirements has inherent limits (providing a rationale for building the NI), but the statistical concept underlying the estimates of undernourishment has been criticised much more heavily. A main argument in this debate is the alleged undue consideration of the variation of dietary energy intakes across households and individuals (Svedberg 1998, Smith 1998). As a side result, this study has demonstrated that the percentage of undernourished is not only significantly more highly correlated with measures of income distribution than GNP per capita, but also outperforms the HDI, the HDI* (which is calculated without the GDP component), a partial NI without the undernourishment component, and even the NI itself in this respect[5], particularly for the income and consumption shares of the poorer quintiles. However, this laudable sensitivity to equidistribution can hardly be credited with the attempted consideration of inequality throughout the estimation procedure of undernourishment. The sensitivity of the undernourishment estimates to distribution mainly results from the fact that dietary energy supply per capita, which forms the main base of these estimates, reflects distributional aspects to a considerable extent (see Table 6.2.1-2 in Chapter 6.2.1). Upper physiological boundaries of dietary energy intake are believed to be responsible for this finding. In fact, the model designed in Chapter 4.1.1 to reproduce the FAO estimates of undernourishment solely with dietary energy supply data (without any explicit information on distribution and average minimum dietary energy requirements) shows that they are predominantly determined by the variation of dietary energy supply per capita. Although the FAO figures on the prevalence of undernourishment may not include as much additional information on distribution as would be desirable, the estimation process obviously implies a sensible transformation function of

[4] This is widely true, although, with respect to anthropometric indicators, Klasen 1999 has raised again the question of possible inadequacies in international reference standards after this debate had seemed to be 'dead'.

[5] On the other hand, the NI is superior to the above-mentioned indicators, including the percentage of undernourished and dietary energy supply per capita, with reference to tight rank correlations with the four most common international poverty measures (see Chapter 6.2.2). A close relation to poverty is considered to be a more important characteristic than sensitivity to distribution, because inequality is of a relative nature, whereas the poverty indicators try to capture absolute deprivation (World Bank 2000).

dietary energy supply per capita (DES), assuming decreasing marginal returns to rising DES for the reduction of undernourishment. Thus, if dietary energy supply per capita is accepted as a useful indicator, there is no compelling reason to discard the percentage of undernourished[6]. However, it cannot be doubted that the validity of the NI would benefit from methodological improvements of assessing undernourishment as they may be achieved in the future by putting more emphasis on food consumption surveys (Smith 1998).

With respect to **comparability over time**, the fact that all three components of the NI are expressed as percentage ratios - thus having natural boundaries - proved advantageous: intertemporal comparability could be preserved due to the omission of standardisation of indicators, whilst avoiding arbitrary setting of upper and lower limits for the partial indicators[7]. Regarding **comparability across countries**, the comparability of the underlying indicators of the NI is essential, and was considered in a conceptual sense by integrating indicators describing distinct physical states (hunger, undernutrition, death) that entail human suffering, irrespective of the cultural context. From a statistical point of view, deficiencies in international comparability may exist for the percentage of undernourished: this has been criticised for ignoring variations in dietary energy requirements across climatic zones (Svedberg 1998). No remedy is available so far but, for the anthropometric indicator, international comparability could be improved by refining the data. Correction factors were computed and, if necessary, applied to the prevalence rates of underweight in children pertaining to other age groups than the under-fives. This was done to adapt the data to a uniform reference age group. Still, limits to the international comparability of the NI may arise from the anthropometric indicator due to small deviations from international reference standards of the growth curves of well-nourished South Asian children (Klasen 1999). The clarification of these issues, however, has to be left to further research.

---

[6] However, one disadvantage of the undernourishment vis-à-vis the dietary energy supply indicator should be mentioned: the admittedly plausible discounting of dietary energy supply per capita in the estimation process of undernourishment implies that small variations in DES cause large fluctuations in the percentage of undernourished, if dietary energy supply per capita is low (see Figure 4.1.1-2 in Chapter 4). Thus, due to this transformation, errors in the assessment of food availability are likely to multiply at low levels of dietary energy– that is, exactly for the poorest countries which usually also have the weakest statistics.

[7] Intertemporal comparability seems to be a feature of international indices that goes without saying. But, as already stated in Chapter 2.1.2, early versions of the HDI failed to fulfil this requirement (see Lüchters and Menkhoff 1994 and Bohnet 1994).

Concerning **ability to differentiate**, a decline can be observed with rising NI values. This is evident from the tendency of differences between NI values for countries with successive ranks to diminish with improving rank positions (compare the international ranking in Table 5.1-1). The NI value for Kuwait on the first rank exceeds the NI for Romania on rank ten by only 1.93 points, which raises the question whether a disparity of ten ranks is justified by the significance of this difference[8]. In contrast, at the bottom end of the ranking, the NI value of Afghanistan on rank 106 is 11.55 points below the NI of Sierra Leone, which is ranked number 96 within the whole set of countries. The comparison of such differences in NI values for equidistant ranks, however, should be seen against the fact that the comparatively wealthy countries at the upper end of the list have more reliable statistical systems at their disposal. Thus, measurement errors in the partial indicators of the NI are expected to be smaller for the top-ranked countries than for the poor and war-torn states at the very bottom of the list.

The general **robustness** of the NI to measurement error cannot be tested empirically because the standard deviations of errors in the underlying data are unknown[9]. Yet, since the data are predominantly taken from different statistical sources and the above mentioned possible deficiencies in measuring the components are unlikely to be related, independently distributed measurement errors are a reasonable assumption, for which a relative decrease in the measurement error of the index was demonstrated in a formal exercise in Chapter 2.2. The requirement that an index should be robust to small changes of the aggregation function is fulfilled for the NI: different NI versions were calculated by using various methods of standardising the components, varying the weights, and replacing the weighted average of first order in the NI formula by the weighted average of second order. These modified versions were shown to be highly redundant vis-à-vis the basic simple NI, with rank correlation coefficients being close to unity (ranging from 0.965 to 0.999) and, in contrast to the findings for the relationship of the NI to its components, similarly high levels of redundancy were found for subsamples of the data set. This property of the NI is rooted in the significant correlations of partial indicators – though they prevent the NI from being completely non-

---

[8] This may underscore the necessity to omit developed countries from NI calculation, because otherwise the ability of the NI to differentiate among the top ranks would further decrease.

[9] A Monte-Carlo simulation based on plausible assumptions about the standard deviations of measurement errors can be run to assess the extent of error in the aggregate variable in such instances. This option, however, was not taken in the present study.

redundant in comparison to its components on the one hand, they lay the ground for its robustness to variations in the aggregation function on the other.

The **stability** of the NI against short-term, meaningless fluctuations can be taken for granted due to the cumulative character of the anthropometric indicator (which considers not only acute, but also chronic undernutrition) and the fact that the percentage of undernourished is based on three-year averages. Regarding the **sensitivity** of the NI to changes, it is assumed to benefit from the inclusion of dietary energy supply besides underweight in children. As has been reported from several case studies, adults are able to protect the nutritional status of their children during transient periods of food deprivation by allocating food preferentially to younger family members[10] (Ferro-Luzzi et al. 2001, Chastre and le Jeune 2001). Yet, a sudden and severe deterioration of food and nutrition security, which may occur due to an epidemic or natural disaster, is likely to find its reflection in increased prevalence of acute undernutrition in children - captured by the underweight indicator - and rises in child mortality. However, it must be admitted that the NI is better suited to indicate long-term trends in the nutrition situation than acute crises, which, moreover, mostly take place within a quite limited geographical area.

8.2 The NI in the Light of the Debate on International Indices

Many concerns and criticisms have been raised with regard to international indices. The following section, which roughly follows the listing of pros and cons in Chapter 2.2 (see Table 2.2-1), mainly discusses the counterarguments with reference to the NI, since the advantages of constructing a Nutrition Index have already been highlighted prior to and in the course of its design. The objection that indices, and international indicators in general, neglect subjective matters applies to the NI also. The NI contains no information on people's chances of meeting their dietary preferences, as claimed by the definition of food security passed at the World Food Summit in 1996 (FAO 1996a). However, in comparison to the debilitating impact of insufficient quantity and quality of food or restrictions in biological utilisation, the consumption of food that does not comply with personal preferences seems to be a minor evil. Actually, in times of food shortage, people resort to eating some kinds of wild foods which they disdain under normal circumstances. Furthermore, although not everything can be captured in

[10] In an Ethiopian case study, Ferro-Luzzi et al. (2001) found a much larger seasonal variation of nutritional status among adults than among children and adolescents.

numbers[11], there is no doubt that quantitative measures of progress are needed for monitoring food security and nutrition - as in other fields, qualitative surveys may complement numerical measures.

Concerning the informative and monitoring function of the NI, it is surely impeded by limited availability and reliability of the underlying data. In comparison to its partial indicators, a disadvantage of the NI (and other indices) consists in the fact that it can only be used for the intersection of countries for which information on all index components is available. The lack to date of anthropometric data for children for a range of countries is particularly limiting. However, since an increasing number of nutrition surveys have been carried out in past years, and data availability at the international level has continuously improved, it is likely that the NI can be calculated for a larger set of countries in the future. Regarding the complaint that indices foster the 'mystification of numbers', it should not be taken too seriously with respect to the Nutrition Index, because variations of the NI over time and across countries can easily be traced back to variations in its components due to its simple construction, offering little scope for mystification[12]. Disregard of the cultural context is no grave accusation against the NI either, because its dimensions relate to very basic human needs and values.

Another argument that has frequently been raised regarding the technical properties of indices, namely that their compilation entails the loss of detailed information, applies only if an index is intended to replace its partial indicators. However, the NI is meant to be a complement to rather than a substitute for existing food security and nutrition indicators. Since the existence of an aggregate measure should keep nobody from looking at the underlying data and related indicators, as well as the broader context, the information contained in the components is not really 'lost'. In fact, the partial indicators of the NI might be given greater attention if the index is able to mobilise political will for improving food and nutrition security by means of programmes and projects tailored to problem areas[13]. Like other indices, the NI may be charged with

---

[11] The quantitative aim agreed upon by the participants at the World Food Summit in 1996 – halving the percentage of undernourished by 2015 (FAO 1996a) – may be regarded as defining a necessary but by no means sufficient condition for more satisfactory nutritional outcomes.

[12] On the other hand, the simplicity and transparency of the NI might be a drawback as the proneness to mystification renders indices more appealing and powerful in the political debate.

[13] The HDI has had such an impact in Egypt: after the release of the HDI, decision-makers realised that the country was doing so badly in terms of human development against the yardstick of GNP per capita because of its weak performance in the education sector. This insight motivated decision-mak-

arbitrary weighting of components. However, the NI was found to be very robust to moderate changes in weighting factors, a characteristic implying that it is not worthwhile discussing weights extensively. It may provide some scope for 'weight pluralism', which means in this case that countries pursuing national priorities deviating from the relative importance attached to NI components by the weights would not be significantly disadvantaged in terms of their ranking position. Anyway, the problem of fixed trade-offs among NI components is not a serious one if the NI is not applied with the intention of unbalanced, 'cost-effective' maximisation (possibly involving gains in one dimension whilst allowing for the degradation of another).

As has already been argued, the computation of the NI is likely to decrease the impact of measurement errors for the total sample of countries. However, a distortion of NI values for a few countries due to spurious data for at least one partial indicator cannot be excluded (the percentage of undernourished for Nigeria and its effect on the respective NI value seems to be such a case, although there may be other reasons for the huge disparity between anthropometric and child mortality data on the one hand and the undernourishment estimate on the other, see Table A 5.1-1). On the other hand, to the extent that the NI promotes a synopsis of food security and nutrition indicators, this may help to detect errors and inconsistencies within the data sets.

Any political impact of the NI is yet to show up, and thus the respective discussion can only be hypothetical. Concerning the criticism that indices are means to impose 'outside priorities' on countries, one should bear in mind that progress in the components constituting the NI is already part of international agreements. Freedom from hunger, adequate nutritional status and favourable survival chances for children are widely accepted goals, and the majority of countries have committed themselves to their promotion. A sense of competition encouraged by the country ranking might indeed foster good policies after a long period of predominant preoccupation with economic growth and/or human development, and a comparative neglect of food security and nutrition. The identification of countries at the bottom of the ranking should not be characterised as 'naming and shaming' or 'intimidation', but act as a signal to the international community and the countries themselves that much more effort and assistance to overcome food insecurity and undernutrition is urgently needed. Whereas indices and indicators used internationally have induced attempts to improve data collec-

ers to redress the bias that existed in favour of university enrolment at the expense of school enrolment, and to take effective measures for reducing the dramatic gender gap in school enrolment rates (Handoussa 2000).

tion as well as to manipulate statistics, it is not yet clear if the NI will bear any such incentives. On the one hand, if widely applied, the NI could reinforce the existing trend of more frequent and comprehensive data collection in the field of food security and nutrition. On the other hand, countries might exaggerate the extent of their nutritional problems in a cry for international assistance, particularly if the NI was used as a criterion for allocating development aid. Other states might falsify statistics to downplay the problems of hunger and malnutrition, and so avoid being blamed for disregard of the most basic human rights. In both cases, surveys carried out by independent agencies and international supervision might help to prevent politically motivated data manipulation.

In conclusion, if the NI was applied on a large scale, this might not only bring the largely unfruitful dispute on measurement issues to an end, but could help to devise more balanced policies aiming simultaneously at ensuring adequate dietary energy supply and creating the preconditions for better biological utilisation of food. Children as a nutritionally vulnerable group might be enabled to transform better access to food more readily into healthy growth and survival. The other way round, child malnutrition and mortality might not only be tackled by investing in health and education, but be increasingly addressed in the context of food supply systems. Nubé (2001), who argues that anthropometric indicators are more reliable for assessing undernutrition than FAO's undernourishment indicator (his evidence is based on a small sample of countries, however, and contradicted by the findings of the present study for a larger set of countries, compare Chapter 4.3), nevertheless acknowledges the relevance of food balance and dietary energy supply data: 'Both methodologies have their own merits and provide different types of information. The Supply Utilisation Accounts, from which the food availability data are derived, provide comprehensive annual information on the flows of agricultural commodities in a country, and are an important source of information for the identification of changes and vulnerabilities in the food supply system' (Nubé 2001, p.1284). The NI seeks to reap benefits from both types of indicators for decision-making. In addition, the risk of bypassing the poorest on an international scale is likely to be lower if the NI is taken as a yardstick for channelling assistance instead of using GNP per capita, the HDI or the index components.

9 Conclusions

Notable progress has been achieved in enhancing food security and preventing malnutrition in the five decades following the Universal Declaration of Human Rights in 1948. Nevertheless, food security and good nutrition for all have remained a distant vision. A world which aims for ending hunger and malnutrition must know how nutrition develops on a country-by–country basis and in large countries also at a subnational level. The new Nutrition Index presented here permits measurement of progress or the lack thereof since the important international conferences, such as the World Summit for Children (1990), the International Conference on Nutrition (1992) and the World Food Summit (1996), where the international community pointed out that the elimination of hunger and malnutrition should be of utmost importance.

A review of international monitoring processes, however, showed that food security and nutrition do not figure very prominently on the development agenda, despite their recognised importance and reconfirmed international commitment to end hunger and malnutrition. Indicators that are widely used in international comparisons – GNP per capita or GDP per capita and the Human Development Index – do not adequately capture food security and nutrition. In addition, different lead indicators are used by FAO, WHO and UNICEF. The lack of a commonly accepted measure for food security and nutrition has to date prevented beneficial synergies that could arise from an integrated approach to nutritional problems.

An international Nutrition Index (NI) has been developed in the present study to create a comprehensive measure of food security and nutrition. In accordance with the theory of international comparisons and a conceptual framework for nutrition, the dimensions of the unobservable abstract variable to be captured were defined as people's ability to be free from hunger, malnutrition and premature death. The following empirical variables were selected to reflect these complementary dimensions of a country's nutrition situation: 1) the percentage of undernourished in the population, 2) the prevalence of underweight in children under the age of five, and 3) the under-five mortality rate, indicating the deadly synergy between inadequate food intake and unhealthy living conditions. Aggregation was carried out by means of weights derived from principal components analysis. Standardisation of partial indicators was omitted, because the calculation of preliminary indices with differently standardised and non-standardised indicators led to highly redundant results, thereby suggesting to opt for the simplest solution. In the final formula, the three indicators are expressed in percentages and aggregated as a linear combination with almost equal weights that is subtracted from 100. Therefore, NI values increase with improvements in food security and nutrition, and the shortfall from 100 corresponds approximately to the average of the proportion of undernourished, underweight pre-school children and child mortality under five referenced to live births (all expressed in %). After data refinement and estimation procedures to supplement data which was lacking, the data available permitted NI calculation for the vast majority of developing countries, as well as for nine countries in transition, and at different points in time (1981, 1992 and 1997).

In order to apply the NI as a monitoring instrument for countries' performance in combating hunger and malnutrition, an international ranking with NI values of 106 developing countries and countries in transition was set up, covering a range of about 43 NI points. Kuwait is at the top of the list with an NI of 98.0, closely followed by Yugoslavia (Serbia and Montenegro), Chile, Libya, Argentina and Lebanon. Countries that experienced long-lasting violent conflicts affecting human capital formation, the infrastructure and the productive base of the economy, have very low NI values, indicating a miserable track record in food security and nutrition. Nine out of the eleven countries at the very bottom of the list – Afghanistan, Ethiopia, Somalia, Eritrea, Mozambique, Democratic Republic of Congo, Angola, Chad and Sierra Leone - were torn by civil wars in the NI reference period, or have only recently recovered from cruel internal conflicts. In 1997, after 18 years without peace, Afghanistan was by far the worst performer in the international comparison of Nutrition Index values, with an NI of only 54.8[1]. Moreover, warfare was frequently accompanied by economic mismanagement such as undue price controls, barriers to internal trade, and market development that were set up by the state. A few examples of countries which suffered such misguided economic policies are Angola, Ethiopia (formerly including Eritrea) and Mozambique.

The graphical illustration of the world nutrition situation in a map showed that countries with bad to extremely bad nutrition situations, indicated by an NI value of less than 80.4[2], are almost exclusively located in Sub-Saharan Africa and South Asia. The calculation of NI values for the states and provinces of the world's two most

---

[1] This NI value arises from a proportion of undernourished of 62% in 1995-97, a prevalence of underweight in children under five of 49.3%, and an under-five mortality rate of 257 per 1000 live births in Afghanistan in 1997.

[2] The mean NI of countries in this group amounts to 71, corresponding to average values of 39% undernourished (which is roughly equivalent to a dietary energy supply of 2000 kcal per capita per day according to the model devised in section 4.1.1) - 34% underweight in children and an under-five mortality rate of 160 per 1000 live births.

populous countries – India and China – underscored the importance of subnational disaggregation: vast disparities in food security and nutrition are found within these countries, expressed in differences of more than ten NI points between best- and worst-performing administrative regions.

As regards NI trends by continents, in Africa as a whole little progress has been made since the beginning of the 1980s, and stagnation is found throughout the 1990s around an NI value of 77. In contrast, Asia has experienced considerable improvements in its overall nutrition situation. Starting with the lowest aggregate NI of 72 in the 1980s, Asia surpassed the African continent in the meantime due to notable reductions of undernourishment, underweight prevalence and child mortality, leading to an eight point gain in NI value. But in 1997, Asia was still far from the achievements reached in Latin America and the Caribbean, where the aggregate NI already amounted to 89 in the beginning of the 1980s. On-going progress can be seen in the latter region, though not at a great pace: the NI increased by about three points from 1981 to 1997.

The disaggregation of data by region disclosed trends which showed that, similarly to Latin America and the Caribbean, North Africa and the Near East experienced small to medium NI improvements from 1981 to 1997. Most Latin American, North African and Middle Eastern countries achieved at least a satisfactory nutrition situation, indicated by NI values greater than 90.5[3] (except for Yemen, having an NI just below 70). The countries in the Caribbean arrived at mixed NI values around 85, but Haiti is still lagging considerably behind with an NI of only 66.5. The few states of Eastern Europe and the former Soviet Union for which data availability permitted NI calculation (Croatia, Yugoslavia, Romania, Russian Federation, Armenia, Azerbaijan, Kazakhstan, Uzbekistan and Kyrgyz Republic) also had good or mixed NI values greater than 85. The food security and nutrition situation was still bad in South Asia and no better than middling in most of South-East Asia, but apart from the depressing situation in Afghanistan, where the NI value deteriorated most from 1981 to 1997, falling by more than 17 points, very promising upward trends were observed for this region. China, India and Nepal, for example, increased their NI values by more than ten points from the beginning of the 1980s to the end of the 1990s. Technological progress in agriculture, economic growth, and an expansion in public service provision helped

---

[3] For the total group of countries with NI values above 90.5, the average NI is 94. The mean values for the NI components are 6.4% for the proportion of undernourished and 7.7% for underweight in children, respectively, and 31 deaths per 1000 live births for the under-five mortality rate.

to bring about large declines in undernourishment and child malnutrition, and to cut child mortality by about one third. In Sub-Saharan Africa, the nutrition situation is bad or even extremely bad in about three quarters of the countries with NI values below 80.4. Some West African states achieved large NI increases over the past two decades, particularly Mauritania, Burkina Faso and Ghana. The considerable change for the better in Ghana – a gain of almost 20 NI points - can be traced back mainly to the favourable impact of market liberalisation and the introduction of high-yielding cassava varieties on domestic staple food production, which benefited dietary energy supply per capita. Child mortality was reduced by one third at the same time, but child malnutrition remained virtually unchanged in Ghana. In large parts of Eastern and Southern Africa the poor nutrition situation at the beginning of the 1980s was deteriorating even until 1997, especially in Tanzania, Zambia, Madagascar, the Democratic Republic of Congo and Angola. Therefore, not only in terms of absolute NI levels, but also with respect to recent trends, Sub-Saharan African countries give rise to the most serious concern about food security and nutrition.

Going beyond a purely descriptive comparison of NI trends and levels across countries and regions, the relationship of the NI to other international indicators was analysed, considering redundancy vis-à-vis indicators with broad data availability, and the ability of the NI to reflect relevant measures with a scanty database. The judgement whether the NI is redundant vis-à-vis other indicators depends on the setting of arbitrary threshold levels for rank correlation coefficients (0.90 for Level 1 and 0.70 for Level 2 redundancy). However, the value added by a new measure such as the NI should not be judged solely from redundancy analysis by means of rank correlations - even in the face of high correlations with other indicators for the total sample, important lessons can be learnt from country cases with large differences in rankings, the 'outliers'.

First, the NI was evaluated against the yardstick of other food security and nutrition indicators. Considering rank correlations with its components for the full sample, Level 1 redundancy was only found for the NI and the percentage of undernourished (the Spearman's correlation coefficient is –0.91), and Level 2 redundancy for underweight in children and under-five mortality (the coefficients are –0.88 and –0.86, respectively)[4]. When the sample was split up into three groups of equal size according to

[4] Since the weighted average of the partial indicators is subtracted from 100 to form the Nutrition Index, the NI and its components are negatively correlated. Therefore, the threshold levels of redundancy are applied accordingly for negative correlation coefficients.

NI level and the correlations were calculated separately for countries with low, medium and high NI, the coefficients fell below the thresholds of redundancy (between –0.24 and –0.67), except for the correlation of the NI with the percentage of undernourished for the low NI group (-0.76). Whereas it may be overly demanding to require non-redundancy of a composite index vis-à-vis its components, the above results demonstrate that the NI adds most to the information contained in its components if subgroups of the full country sample are considered. In a further step, the analysis of rank correlations was extended to indicators of micronutrient deficiencies that were not included in the index due to scanty data availability. In this case, high rank correlations are desirable because they indicate the ability of the NI to reflect nutritional information which had not been included. Compared to its components, the NI performed quite well in terms of correlations with the goitre rate (indicating iodine deficiency), anaemia (reflecting iron deficiency), and both subclinical and clinical Vitamin A deficiency. The coefficients referring to the NI all exceeded the lowest coefficients found for its partial indicators and the micronutrient deficiencies considered, and in two cases came very close to the highest coefficients pertaining to the single indicators. A notably high correlation exists between the NI and severe Vitamin A deficiency (Spearman rank correlation coefficient of –0.71 for the small sample of 29 countries with available data). The close relationship between this deficiency and under-five mortality as an NI component can be held responsible for this. Last, but not least, rank correlations were calculated for the NI and the Hunger Index (HI) devised by Douglas Hicks as a comprehensive measure of food security. The sample under consideration comprised only 48 countries due to limited data availability for the HI. Given the small number of cases, the correlation coefficient of 0.91 is quite high and indicates Level 1 redundancy. If the sample was divided into two groups according to NI level, the coefficients still indicated Level 2 redundancy (0.81 and 0.78 for lower and higher NI values, respectively). Though the NI and the HI are quite different in their design, both indices thus lead to very similar rankings. Yet the NI is superior to the HI with respect to data availability.

Second, the frequently voiced critique that indicators referring to national aggregates conceal inequality and convey little information on poverty was taken into account by exploring the sensitivity of the NI to distribution and income poverty. Corresponding to similar results for the HDI in other studies, the analysis demonstrated that the NI bears an 'equidistributional bias', which means that it is significantly more sensitive to inequality than GNP per capita (distribution is measured by the Gini coefficient, the ratio of the richest to the poorest quintile, and shares of income or consumption quintiles). The NI is also slightly superior to the HDI in this respect, but the relationship of ranking differences between GNP per capita and the undernourishment component of the NI to distribution was found to be even more pronounced.

Furthermore, the NI showed significantly higher correlations with four common measures of absolute poverty (proportion of the population below 1$ a day, poverty gap at 1$ a day and the two respective indicators for the poverty line at 2$ a day) than did GNP per capita and the HDI. The fact that the NI is more sensitive to income poverty than currently prevailing development indicators, and is even more sensitive in this respect than the single indicators it is composed of, can be considered a noteworthy comparative advantage of the newly created index. This holds especially true if targeting in the field of food security and nutrition is to be in line with the needs of the poorest, as has often been claimed. In comparison, the poverty measures themselves are of limited usefulness for poverty-sensitive targeting due to limited data availability: the NI has a markedly wider coverage than explicit data on absolute poverty (it is available for 106 countries in the 1990s, 46 of which are without recent information on monetary poverty according to international poverty lines[5]). Furthermore, income poverty measures neglect the availability of basic public goods and their role for standards of living, whereas non-monetary elements of poverty are considered in the NI via the impact of public services on child nutrition and survival. Taking the innovative measure of the NI seriously instead of aiming solely at maximisation of GNP or HDI would compel decision-makers to address poverty more effectively.

Third, the relationship of the NI to international development indicators was investigated. For the NI and GNP per capita, the Human Development Index (HDI) and life expectancy, which is the component of the HDI meant to reflect health and nutrition, the Spearman rank correlation coefficients varied between 0.82 and 0.87, indicating Level 2 redundancy. Regarding the correlation coefficients for subsamples of equal size with low, medium and high NI values, they all fell far below the lower threshold level of 0.70 (being in the range of 0.20 to 0.53 and partly insignificant). Again, the information gained by using the NI is highest for subgroups of the full country sample. A closer investigation of the relationship between GNP per capita and the NI revealed a logarithmic functional form. This means that large improvements in

[5] The robustness of the results suggests that the close relationship of the NI to the poverty measures also holds for the large set of countries not included in the analysis due to lack of poverty data.

food security and nutrition can be brought about most easily by rises in economic performance if the initial GNP per capita of a country is low. The bulk of African countries were doing poorly with reference to both indicators, which is attributable to the situation in Sub-Saharan Africa, where 30 out of 43 countries with available data had NI values below 80 and their average GNP per capita was less than 1000 PPP$. In contrast, the majority of Asian and Latin American countries were found in a higher income class, with a GNP per capita of between 1500 and 7000 PPP$, and better NI scores. Moreover, on the African continent, quite a large number of countries had considerably worse NI values than would be expected from their GNP per capita – this applied in particular to the Southern African countries Namibia, Angola and Botswana, to Eritrea and Ethiopia in East Africa and to Niger in West Africa. Their NI values are more than eight points below the values predicted according to GNP per capita. In contrast, the nutrition situation in Benin, in the North African states of Morocco and Egypt, in Senegal and in Tanzania is better than expected according to GNP per capita. In Asia, China and some countries in the Near East managed to convert economic resources quite well into nutritional well-being for their people. Substantially lower NI scores than predicted on the basis of GNP per capita could be observed for Thailand, Cambodia and Bangladesh (eight points and more below predicted NI values). In Latin America and the Caribbean, only Jamaica, with a quite good nutrition situation from the viewpoint of its economic development, and Haiti, with a considerably lower NI than predicted (more than ten points less), can be considered as outliers.

To analyse reasons for outliers more thoroughly, four pairs of low-income countries from three continents and with almost equal GNP per capita for each pair, but large disparities in NI values, between 8 and 18 points, were selected. Their situation was scrutinised for explanations of positive or negative deviations of the NI from the values predicted according to GNP per capita. Countries that are negative outliers were more frequently found to be characterised by an irresponsible elite, protracted warfare and political instability, neglect of the health and education sector, and undue restrictions on agricultural markets, sometimes in combination with natural calamities like drought. These characteristics pertain to Ethiopia and Angola, and partly to Haiti. In contrast, positive outliers mostly promoted market incentives in agriculture (e.g., Benin), demonstrated strong commitment to basic education and primary health care (for example Tanzania and Nicaragua), and were less affected by violent conflicts, coups d'état and natural disasters.

Outliers are also identified with respect to the HDI. In contrast to GNP per capita, the HDI includes health and education variables relating to nutrition and child survival, but it likewise fails to capture disparities in food supply among countries. Therefore, despite a relatively tight correlation between the two indices, large divergences in NI values of 15 points and more can be found for countries at comparable stages of 'human development'.

A more systematic empirical analysis of NI variation in a multivariate regression confirmed that national availability of economic resources, indicated by GNP per capita, is the main determinant of food security and nutrition as measured by the NI: the logarithm of GNP per capita alone accounted for 74% of NI variation. Mismanaged economies have deadly consequences, and economic growth is central to the amelioration of desolate nutrition situations[6]. To achieve this purpose best, growth needs to be broad-based and to contribute to poverty alleviation. However, there remains a lot of scope for policies to relieve hunger and malnutrition in many countries, independent of the pace of GNP growth. This is illustrated by the fact that the adjusted coefficient of determination can be raised to 0.876 if political and social indicators are entered as explanatory variables in the regression of the NI. Contrary to expectations, *ceteris paribus*, low degrees of political freedom seem to be associated with relatively higher NI values. Therefore, the presumption that democratisation can counteract hunger is not supported by the findings of the present study. However, high levels of military expenditure and warfare are clearly detrimental to food security and nutrition, even apart from their impact on macro-economic performance. Countries that were engaged in a war in the NI reference year, or in at least one of the nine preceding years, had about three points lower NI values than countries experiencing peace throughout that period – a significant difference that is not accounted for by other economic, political or social variables or regional dummies in the model. Preliminary evidence suggested considerable lagged effects of violent conflicts, and long-lasting wars proved more destructive than short episodes of fighting, irrespective of their effect on GNP per capita. More attention should, therefore, be given to conflict prevention and resolution as well as to rehabilitation measures in the field of food security and nutrition after peace

[6] According to the coefficient gained in the multivariate regression – which gives only the partial effect of changes in GNP per capita – an increase in one unit of logarithmic GNP per capita raises the NI by 6.7 points. One unit gain in logarithmic GNP per capita is equivalent to adding 860 PPP$ to an initial level of 500 PPP$ (which is about the lowest level of GNP per capita found in the data set). However, if a higher income of 3000 PPP$ is the starting point, one unit increase in the logarithm corresponds to raising GNP per capita by around 5150 PPP$.

has been restored. As long as armed conflicts exist, 'food security for all will remain a dream' (Guiton 2001, p.2). Not surprisingly, engagement of countries in wars was shown to be associated with relatively bad governance. Consequently, no significant impact of the proxy for the quality of governance (the ICRG index) on food security and nutrition could be found in the presence of a variable relating to warfare. Interestingly, the significance of the ICRG index, when used in the absence of a war variable, could only be traced back to the components of the index associated with a climate of legal and economic insecurity ('rule of law', 'expropriation risk', 'repudiation of contracts by the government'), but not to the two subindices indicating weak and corrupt bureaucracies. Yet with respect to the political variables, it should be noted that the estimation strategy identifies partial rather than total effects of political freedom, good governance or warfare, which may work just as well through other variables in the model (e.g., GNP per capita and indicators of public service provision). The results should, therefore, not provoke an underestimation of the benefits of peace-building and strengthening institutions for nutritional well-being.

A climate of peace and security can create the preconditions for using public resources wisely and effectively for necessary investments in the agricultural, health and education sectors. Agriculture can play a key role in fostering broad-based economic development and in improving food supply, especially in poor countries. As regards health and education, a high degree of urbanisation[7] turns out to be favourable to child nutrition and survival. A difference of 10% in the share of the population living in urban areas is associated with a divergence in NI values of 1.4 points. With reference to health care, encouraging actual utilisation of services is more promising than merely expanding physical access. This showed up in the effects of immunising children against DPT and the role of births attended by health staff. Both utilisation rates relate to the effectiveness of the health care system, hinging, among other factors, on the knowledge and compliance of the population. Furthermore, they may be, at least to some extent, indicative of the status of women and children in society. To enhance utilisation of these services, not only physical access to health facilities, but also an understanding of the benefits of care, the opportunity and the willingness of potential beneficiaries to use them should be promoted.

---

[7] Urbanisation is considered as a proxy for knowledge and health environments, because access to basic facilities and quality health care, educational opportunities and incentives are generally better in urban than in rural areas.

In conclusion, the improvement of food security and nutrition remains very much a task of public action by both government and civil society organisations. Well-designed, effective food- and nutrition-oriented policies contribute directly to people's welfare, but do so indirectly also by raising their capacity for work and their incomes. International support for investment in the agriculture, food, health and education sectors is needed and can produce high returns, if not counteracted by bad governance and military conflict. The NI could thus also serve as an important tool for monitoring a country's performance when receiving or not receiving debt relief or incremental development aid. In its function as an international monitoring tool, it is hoped that the Nutrition Index will make both developing and developed countries, national and international players, more accountable for their commitments and will help to speed up urgently needed improvements in food security and nutrition.

10 Summary

Notable progress has been achieved in enhancing food security and preventing malnutrition in the five decades following the Universal Declaration of Human Rights in 1948. Nevertheless, food security and good nutrition for all have remained a distant vision. A review of international monitoring processes showed that food security and nutrition do not figure very prominently on the development agenda despite their recognised importance and the reconfirmed commitment to end hunger and malnutrition. Moreover, the lack of a commonly accepted measure for food security and nutrition has to date prevented beneficial synergies that could arise from a better integrated approach to nutritional problems.

An international Nutrition Index (NI) has been developed in the present study to create a comprehensive measure of food security and nutrition. The following empirical variables were selected to reflect complementary dimensions of a country's nutrition situation: 1) the percentage of undernourished in the population, 2) the prevalence of underweight in children under the age of five and 3) the under-five mortality rate. Aggregation was carried out by means of weights derived from principal components analysis. The three indicators are expressed in percentages and aggregated as a linear combination that is subtracted from 100. Therefore, NI values increase with improvements in food security and nutrition. After data refinement and estimation procedures to supplement data which was lacking, the data available permitted NI calculation for the vast majority of developing countries, as well as for nine countries in transition, and at different points in time (1981, 1992 and 1997).

In order to apply the NI as a monitoring instrument for countries' performance in combating hunger and malnutrition, an international ranking with NI values of 106 countries (including 97 developing countries and nine countries in transition) was set up. Kuwait is at the top of the list with an NI of 98.0, closely followed by Yugoslavia (Serbia and Montenegro), Chile, Libya, Argentina and Lebanon. Nine out of the eleven countries at the very bottom of the list – Afghanistan, Ethiopia, Somalia, Eritrea, Mozambique, Democratic Republic of Congo, Angola, Chad and Sierra Leone - were torn by civil wars in the NI reference period, or have only recently recovered from cruel internal conflicts. In 1997, after 18 years without peace, Afghanistan was by far the worst performer in the international comparison of Nutrition Index values, with an NI of only 54.8. Moreover, warfare was frequently accompanied by economic mismanagement such as undue price controls and barriers to internal trade that were set up by the state (examples are Angola, Ethiopia - formerly including Eritrea - and Mozambique).

The graphical illustration of the world nutrition situation in a map showed that countries with a bad to extremely bad nutrition situation, indicated by an NI value of less than 80.4[1], are almost exclusively located in Sub-Saharan Africa and South Asia. The calculation of NI values for the states and provinces of the world's two most populous countries – India and China – underscored the importance of subnational disaggregation: differences of more than ten NI points between best- and worst-performing administrative regions are found within these countries.

Most Latin American, North African and Middle Eastern countries achieved at least a satisfactory nutrition situation. The countries in the Caribbean arrived at mixed NI values around 85; however, Haiti is still lagging considerably behind with an NI of only 66.5. The few states of Eastern Europe and the former Soviet Union for which data availability permitted NI calculation also had good or mixed NI values greater than 85. The food security and nutrition situation was still bad in South Asia and no better than middling in most of South-East Asia, but, apart from Afghanistan which holds the world-wide record for the largest fall in NI values 1981 to 1997 (-17.9 points), very promising upward trends were observed for this region. In Sub-Saharan Africa, the nutrition situation is bad or even extremely bad in about three-quarters of the countries, where NI values are below 80.4. Some West African states achieved large NI increases over the past two decades, particularly Mauritania, Burkina Faso and Ghana. In large parts of Eastern and Southern Africa the poor nutrition situation at the beginning of the 1980s was deteriorating even until 1997. Therefore, not only in terms of absolute NI levels, but also with respect to recent trends, Sub-Saharan African countries give rise to the most serious concern about food security and nutrition.

Going beyond a purely descriptive comparison of NI trends and levels, the relationship of the NI to other international indicators was analysed. Considering rank correlations with its components for the full sample, Level 1 redundancy was only found for the NI and the percentage of undernourished[2]. When the sample was split up into

[1] The mean NI of countries in this group amounts to 71, corresponding to average values of 39% undernourished (which is roughly equivalent to a dietary energy supply of 2000 kcal per capita per day according to the model devised in section 4.1.1), 34% underweight in children and an under-five mortality rate of 160 per 1000 live births.

[2] Since the weighted average of the partial indicators is subtracted from 100 to form the Nutrition Index, the NI and its components are negatively correlated. Therefore, the threshold levels of redun-

three groups of equal size according to NI level and the correlations were calculated separately for countries with low, medium and high NI, the coefficients (ranging from –0.24 to –0.67) fell below the thresholds of redundancy, except for the correlation of the NI with the percentage of undernourished for the low NI group (-0.76). Whereas it may be overly demanding to require non-redundancy of a composite index vis-à-vis its components, the above results demonstrate that the NI adds most to the information contained in its components if subgroups of the full country sample are considered. Additionally, rank correlations of the NI and indicators of micronutrient deficiencies, which were not included in the index due to scanty data availability, were determined[3]. Compared to its components, the NI performed quite well in terms of correlations with the goitre rate (indicating iodine deficiency), anaemia (reflecting iron deficiency), and both subclinical and clinical Vitamin A deficiency. Rank correlations were also calculated for the NI and the Hunger Index (HI) devised by Douglas Hicks as a comprehensive measure of food security. Though the NI and the HI are quite different in their design, both indices lead to largely redundant rankings, yet the NI has much broader data availability.

The sensitivity of the NI to distribution and income poverty was also explored by means of rank correlations. Corresponding to similar results for the HDI in other studies, the analysis demonstrated that the NI bears an 'equidistributional bias', which means that it is significantly more sensitive to inequality than GNP per capita. Furthermore, the NI showed significantly higher correlations with four common measures of absolute poverty (proportion of the population below 1$ a day, poverty gap at 1$ a day and the two respective indicators for the poverty line at 2$ a day) than did GNP per capita and the HDI. The fact that the NI is more sensitive to income poverty than currently prevailing development indicators, and is even more sensitive in this respect than the single indicators it is composed of, can be considered a noteworthy comparative advantage of the newly created index. Taking the innovative measure of the NI seriously instead of aiming solely at maximisation of GNP or HDI would compel decision-makers to address poverty more effectively.

The relationship of the NI to international development indicators was likewise investigated. For the NI and GNP per capita, the Human Development Index (HDI)

---

dancy – 0.90 for Level 1 redundancy and 0.70 for Level 2 redundancy - are applied accordingly for negative correlation coefficients.

[3] In this case, high rank correlations are desirable because they indicate the ability of the NI to reflect nutritional information which has not been included.

and life expectancy, the Spearman rank correlation coefficients varied between 0.82 and 0.87, indicating Level 2 redundancy. Again, the information gained by using the NI is much higher for subgroups of the full country sample. The relationship between GNP per capita and the NI was found to be logarithmic, which implies decreasing marginal returns of economic growth to food security and nutrition. On the African continent, quite a large number of countries had considerably worse NI values than would be expected from their GNP per capita, e.g. Namibia, Angola, Botswana, Eritrea, Ethiopia and Niger. The nutrition situation in Benin, Morocco, Egypt, Senegal and Tanzania is better than predicted according to GNP per capita. In Asia, China and some countries in the Near East managed to convert economic resources quite well into nutritional well-being for their people. Thailand, Cambodia and Bangladesh were performing relatively badly in comparison to their GNP per capita. In Latin America and the Caribbean, only Jamaica, with a quite good nutrition situation from the viewpoint of its economic development, and Haiti, with a considerably lower NI than predicted, can be considered as outliers. In an analysis of four pairs of low-income countries, countries that are negative outliers were more frequently found to be characterised by an irresponsible elite, protracted warfare and political instability, neglect of the health and education sector, and undue restrictions on agricultural markets, sometimes in combination with natural calamities like drought (Ethiopia, Angola, Haiti). In contrast, positive outliers mostly promoted market incentives in agriculture (e.g. Benin), demonstrated strong commitment to basic education and primary health care (for example Tanzania and Nicaragua), and were less affected by violent conflicts, coups d'état and natural disasters.

A multivariate analysis revealed that national availability of economic resources, indicated by GNP per capita, is the main determinant of food security and nutrition as measured by the NI: the logarithm of GNP per capita alone accounted for 74% of NI variation. However, there remains a lot of scope for policies to relieve hunger and malnutrition in many countries, independent of the pace of GNP growth. This is illustrated by the fact that the adjusted coefficient of determination can be raised to 0.876 if political and social indicators are entered as explanatory variables in the regression. Contrary to expectations, *ceteris paribus*, low degrees of political freedom seem to be associated with relatively higher NI values. However, high levels of military expenditure and warfare are clearly detrimental to food security and nutrition, even apart from their impact on macro-economic performance. Countries that were engaged in a war in the NI reference year or in one or more of the nine preceding years, had about three points lower NI values than countries experiencing peace throughout that period – a significant difference that is not accounted for by other economic, political or social variables or regional dummies in the model. Not surprisingly, engagement of countries in wars was shown to be associated with relatively bad governance. Consequently, no significant impact of the proxy for the quality of governance (the ICRG index) on food security and nutrition could be found in the presence of a variable relating to warfare[4]. A climate of peace and security can create the preconditions for using public resources wisely and effectively for necessary investments in the agricultural, health and education sectors. Agriculture can play a key role in fostering broad-based economic development and in improving food supply, especially in poor countries. As regards health and education, a high degree of urbanisation[5] turned out to be favourable to child nutrition and survival. With reference to health care, encouraging actual utilisation of services is more promising than merely expanding physical access. This showed up in the effects of immunising children against DPT and the role of births attended by health staff. Both utilisation rates relate to the effectiveness of the health care system, hinging, among other factors, on the knowledge and compliance of the population.

International support for investment in the agriculture, food, health and education sectors is needed and can produce high returns, if not counteracted by bad governance and military conflict. It is hoped that the Nutrition Index will make both developing and developed countries, national and international players, more accountable for their commitments and will help to speed up urgently needed improvements in food security and nutrition.

---

[4] With respect to the political variables, it should be noted that the estimation strategy identifies partial rather than total effects of democracy, good governance or warfare, which may work just as well through other variables in the model (e.g. GNP per capita and indicators of public service provision).

[5] Urbanisation is considered as a proxy for knowledge and health environments, because access to basic facilities and quality health care, educational opportunities and incentives are generally better in urban than in rural areas.

References

Adelman, Irma; Robinson, Sherman (1989), *Income distribution and development*. In: Chenery, Hollis B.; Srinivasan, Thirukodikaval N. (eds.), *Handbook of Development Economics*, Vol. II. North-Holland, Amsterdam.

Agarwal, Bina (1997), *'Bargaining' and Gender Relations: Within and beyond the Household*. Food Consumption and Nutrition Division (FCND) Discussion Paper No. 27. IFPRI, Washington, D.C..

Agung, Igusti Ngurah (1996), The Development of Composite Indices for the Quality of Life and Human Resources Using Factor Analysis. *Journal of Population* 2, No. 2, 207-217.

Ahmed, Akhter U.; Bouis, Howarth E.; Gutner, Tamar; Löfgren, Hans (2001), *The Egyptian Food Subsidy System: Structure, Performance, and Options for Reform*. IFPRI Research Report No. 119. IFPRI, Washington, D. C..

Ahmed, Raisuddin; Haggblade, Steven (2000), *Introduction*. In: Ahmed, Raisuddin; Haggblade, Steven; Chowdhury, Tawfiq-e-Elahi (eds.), *Out of the Shadow of Famine – Evolving Food Markets and Food Policy in Bangladesh*. Johns Hopkins University Press, Baltimore.

Alexandratos, Nicholas (ed.) (1995), *World agriculture: towards 2010; an FAO study*. Wiley, Chichester.

Anand, Sudhir; Ravallion, Martin (1993), Human development in poor countries: on the role of private incomes and public services. *Journal of economic perspectives* 7, No. 1, 133-150.

Azam, Jean-Paul; Collier, Paul; Cravinho, Andrea (1994), Crop Sales, Shortages and Peasant Portfolio Behaviour: An Analysis of Angola. *The Journal of Development Studies* 30, No. 2, 361-379.

Bailey, K.; de Onis, Mercedes; Blössner, Monika (1998), *Protein-Energy Malnutrition*. In: Murray C. J. L.; Lopez, A. D. (eds.), *Malnutrition and the Burden of Disease: The Global Epidemiology of Protein-Energy Malnutrition, Anemias and Vitamin Deficiencies*. The Global Burden of Disease and Injury Series Vol. 8, Cambridge.

Baquet, Sophie; van Herp, Michelle (2000), A Pellagra epidemic in Kuito, Angola. *Field Exchange* 10, (online journal). Posted at http://www.ennonline.net/fex/.

Bardhan, Pranab (1982), Little Girls and Death in India. *Economic and Political Weekly*, September 1982, 1448-1450.

Barth, Christian A.; Karst, Heinz; Petzke, Klaus J. (1995), *Untergewicht und Hungerstoffwechsel*. In: Biesalski, Hans-Konrad; Fürst, Peter; Kasper, Heinrich; Kluthe, Reinhold; Pölert, Wolfgang; Puchstein, Christian; Stähelin, Hannes B. (eds.), *Ernährungsmedizin*. Thieme Verlag, Stuttgart.

Batusa, Rebecca J. (1999), *Lessons Learned with Dairy Cooperative Based Health Insurance Programs in Uganda*. Paper presented at the PHR Study Design Workshop in Mombasa, January 1999.

Berry, La Verle (ed.) (1994), *Ghana – a country study*. Federal Research Division, Library of Congress, Washington, D. C.

Biesalski, Hans-Konrad (1995), *Vitamine*. In: Biesalski, Hans-Konrad; Fürst, Peter; Kasper, Heinrich; Kluthe, Reinhold; Pölert, Wolfgang; Puchstein, Christian; Stähelin, Hannes B. (eds.), *Ernährungsmedizin*. Thieme Verlag, Stuttgart.

Bleymüller, Josef; Gehlert, Günther; Gülicher, Herbert (1994), *Statistik für Wirtschaftswissenschaftler*. 9th edition. Vahlen, München.

Bohnet, Frank (1994), *Der Human Development Index des UNDP – eine kritische Analyse*. Arbeitspapier No. 82, Technische Hochschule Darmstadt, Institut für Volkswirtschaftslehre.

Bread for the World Institute (2001), *Foreign Aid to End Hunger. Hunger 2001*. Eleventh Annual Report on the State of World Hunger. Bread for the World Institute, Washington, D.C..

Bucagu, Maurice (2000), *Mouvements de populations et changements démographiques au Rwanda depuis 1994*. Paper presented at the IAOS 2000 conference on 'Statistics, Human Rights and Development', Montreux, September 4-8, 2000.

Castles, Ian (2000), *The Human Development Report 1997*. Room document submitted to the thirty-first session of the United Nations Statistical Committee. Quoted in Szilágyi 2000.

Census Data Online (2001), *Census of India*. Posted at http://www.censusindia.net.

Central Statistical Organization, Sana'a, Yemen (1998), *Yemen democratic and maternal and child health survey 1997*. Demographic and Health Surveys. Sana'a, Yemen

Chastre, Claire; le Jeune, Sonya (2001), Strengthening analysis of the nutrition situation through linking food security and nutrition information: pitfalls and potentials. *Field Exchange* 12 (online journal). Posted at http://www.ennonline.net/fex/.

Chunming, Chen; Zongming, Shao (eds.) (1994), *Food, Nutrition and Health Status of Chinese in Seven Provinces 1990*. China Statistical Publishing House, Beijing.

Criel, Bart (1998a), *District-based Health Insurance in sub-Saharan Africa*. Part II: Case-studies. Studies in Health Services Organisation and Policy 10, Antwerp.

Criel, Bart (1998b), *District-based Health Insurance in sub-Saharan Africa*. Part I: From Theory to Practice. Studies in Health Services Organisation and Policy 9, Antwerp.

Dasgupta, Partha S.; Weale, Martin R. (1992), On measuring the quality of life. *World Development* 20, No. 1, 119-131.

De Gruyter (ed.) (1998), *Pschyrembel Klinisches Wörterbuch*. De Gruyter, Berlin.

De Silva, Gamini S.; Thattil, S.; Gamini, Samita (2000), *Construction of a Composite Index of Human Development for Developing Nations*. Paper presented at the IAOS 2000 conference on 'Statistics, Human Rights and Development', Montreux, September 4-8, 2000.

Desai, Meghnad (1991), Human Development – Concepts and Measurement. *European Economic Review* 35, 350-357.

Directorate of Economics and Statistics (1995), *Indian Agriculture in Brief.* 25th Edition. Government of India Press, New Delhi.

Drèze, Jean; Sen, Amartya K. (eds.) (1989), *Hunger and public action.* Clarendon Press, Oxford.

EIU (1999), *Country Profile: Nicaragua, Honduras.* EIU, London.

Elmadfa, Ibrahim; Aign, Waltraute; Muskat, Erich; Fritzsche, Doris; Cremer, Hans-Dietrich (1988), *Die große GU-Nährwert-Tabelle: Kalorien/Joule- und Nährstoffgehalte unserer Lebensmittel.* 2nd edition. Gräfe und Unzer, München.

Engle, Patrice L.; Menon, Purnima; Haddad, Lawrence (1999), Care and Nutrition: Concepts and Measurement. *World Development* 27, No. 8, 1309-1337.

Fachredaktion des Bibliographischen Instituts (ed.) (1985), *Duden Rechnen und Mathematik.* 4th edition. Bibliographisches Institut, Mannheim; Wien; Zürich.

FAO (1996a), *Rome Declaration on World Food Security and World Food Summit Plan of Action.* FAO, Rome.

FAO (1996b), *The Sixth World Food Survey.* FAO, Rome.

FAO (1999a), *The State of Food Insecurity in the World 1999.* FAO, Rome.

FAO (1999b), *The FAOSTAT Database.* Posted at http://apps.fao.org/cgi-bin/nph-db.pl. (Accessed in 1999 for data on dietary energy supply per capita).

FAO (2000a), *The State of Food Insecurity in the World 2000.* FAO, Rome.

FAO (2000b), *The FAOSTAT Database.* Posted at http://apps.fao.org/cgi-bin/nph-db.pl. (Accessed in 2000 for data on roots and tubers production and Food Balance Sheet of China).

FAO (2001a), *The State of Food Insecurity in the World 2001.* FAO, Rome.

FAO (2001b), *AIDS – a threat to rural Africa.* FAO Focus Paper Posted at http://www.fao.org/Focus/E/aids/aids1-e.htm.

FAO (2002), *Afghanistan's women – the hidden strength of a war-torn land.* News and Highlights. FAO, Rome. Posted at http://www.fao.org/news/2002/020105-e.htm.

FAO/GIEWS (2001), *The GeoWeb System.* Posted at http://geoweb.fao.org.

FAO/WHO (1992), *World Declaration and Plan of Action for Nutrition.* Rome, December 1992.

Feldbrügge, Torsten (2000), *Economic Aspects of Disaster Management in Developing Countries.* Peter Lang, Köln.

Ferro-Luzzi, Anna; Morris, Saul S., Taffesse, Samson; Demissie, Tsegaye; D'Amato, Maurizio (2001), *Seasonal Undernutrition in Rural Ethiopia: Magnitude, Correlates, and Functional Significance.* IFPRI Research Report No. 118. IFPRI, Washington, D.C.

Frankenberger, Timothy R.; Oshaug, Arne; Smith, Lisa C. (1997), *A definition of nutrition security*. CARE Mimeo. Quoted in Smith (1998).

Freedomhouse (2000), *Freedomhouse country ratings*. Annual updates are posted at http://www.freedomhouse.org/ratings/index.htm.

Frongillo, Edward A.; de Onis, Mercedes; Hanson, K. M. P. (1997), Socioeconomic and demographic factors are associated with worldwide patterns of stunting and wasting of children. *Journal of Nutrition* 127, December, 2302-2309.

Gabbert, Silke (2000), *Ernährungssicherung durch Nahrungsmittelhilfe? Eine Analyse der Vergabeeffektivität ausgewählter Geber*. Agrimedia, Bergen/Dumme.

Garrett, James L. (2001), *Achieving Urban Food and Nutrition Security in the Developing World*. In: Pinstrup-Andersen, Per; Pandya-Lorch, Rajul (eds.), *The Unfinished Agenda – Perspectives on Overcoming Hunger, Poverty, and Environmental Degradation*. IFPRI, Washington, D. C..

Gibbons, Elizabeth D. (1999), *Sanctions in Haiti: Human Rights and Democracy under Assault*. The Washington Papers 177 (published with the Center for Strategic and International Studies, Washington), Praeger Publishers, Westport.

Gillespie, Stuart; Mason, J.; Martorell, Reynaldo (1996), *How nutrition improves*. ACC/SCN Nutrition Policy Discussion Paper No. 15. UN ACC/SCN, Geneva.

Gilson, Lucy (1998), *The Lessons of User Fee Experience in Africa*. In: Beattie, A. et al. (eds.), *Sustainable Health Care Financing in Southern Africa*. Papers from an EDI Health Policy Seminar held in Johannesburg, South Africa, June 1996. EDI Learning Resources Series, Washington, D.C.

Green, Reginald (1994), *The Course of the Four Horsemen: The Cost of War and its Aftermath in Sub-Saharan Africa*. In: Macrae, Joanna; Zwi, Anthony (eds.), *War and Hunger: Rethinking International Response in Complex Emergencies*. Zed, London.

Guédémé, Alphonse, Houndékon, Rafatou (1997), *Santé de la Mère et de l'Enfant*. In: Kodjogbé, N.; Mboup, G.; Tossou, J.; de Souza, L.; Gandaho, T.; Guédémé, A.; Houedokoho, T.; Houndékon, R.; Tohouegnon, T.; Zomahoun, S.; Capo-Chichi, V.; Cossi, A.: *Bénin - Enquête Démographique et de Santé 1996*. Macro International Inc., Calverton, Maryland.

Guiton, Philippe (2001), *Conflict and Hunger*. Paper presented at the 2020 Vision Conference in Bonn, September 4-6, 2001.

Gujarati, Damodar N. (1995), *Basic Econometrics*. MacGraw-Hill, Singapore.

Haddad, Lawrence (2001), *Burying Nutrition Myths and Activating Choices for Our Children's Development*. Paper presented at the 2020 Vision Conference in Bonn, September 4-6, 2001.

Haddad, Lawrence; Bhattarai, Saroj; Immink, Maarten; Kumar, Shubh (1998), Estimating the interactions between household food security and preschool diarrhea. *Food Policy* 23, No. 3/4, 241-261.

Haddad, Lawrence; Hoddinott, John; Alderman, Harold (1994), *Intrahousehold resource allocation: an overview*. Policy research working paper 1255. Poverty and Human Resources Division. World Bank, Washington, D.C.

Haggerty, Richard A. (eds.) (1989), *Haiti: a country study*. Federal Research Division, Library of Congress, Washington, D.C.

Handoussa, Heba (2000), *Human Development: Design of Indicators: Comment*. Paper presented at the IAOS 2000 conference on 'Statistics, Human Rights and Development', Montreux, September 4-8, 2000.

Hansen, Gerd (1993), *Quantitative Wirtschaftsforschung*. Vahlen, München.

Heltberg, Rasmus; Johannesen, Niels (2001), *How Parental Education Affects Child Human Capital: Evidence from Mozambique*. Unpublished Manuscript, Institute of Economics, University of Copenhagen.

Hicks, Douglas A. (1997), The Inequality-Adjusted Human Development Index: A Constructive Proposal. *World Development* 25, No. 8, 1283-1298.

HIIK (2000), *The KOSIMO database*. Posted at http://www.hiik.de/en/kosimo/kosimo.htm.

Hopkins, Michael (1991), Human Development Revisited: A New UNDP Report. *World Development* 19, No. 10, 1469-1473.

Horton, Susan (1999), Opportunities for Investments in Nutrition in Low-Income Asia. *Asian Development Review* 17, No. 1/2, 246-273.

IAOS (2000), *Conference on 'Statistics, Human Rights and Development'*, Montreux, September 4-8, 2000. Proceedings are posted at http://www.statistik.admin.ch/about/international/03iaos_det.htm.

Icart, Eric; Trapp, James N. (1999), Agricultural and Food Price Policy in Haiti. *Social and Economic Studies* 48, No. 3, 151-167.

IFPRI (2001), *Sustainable Food Security for All by 2020*. Draft Paper for the 2020 Vision Conference in Bonn, September 4-6, 2001. IFPRI, Washington, D.C.. Posted at http:// www.ifpri.org.

IMF (1997), *Angola - Recent Economic Developments*. IMF Staff Country Report No. 97/112, IMF, Washington, D.C.

IMF (1999a), *Angola: Statistical Annex*. IMF Staff Country Report No. 99/25, IMF, Washington, D.C.

IMF (1999b), *Haiti: Staff Report for the 1999 Article IV Consultation*. IMF Staff Country Report No. 99/105, IMF, Washington, D.C..

IMF (1999c), *Nicaragua: Recent Economic Developments and Statistical Annex*. IMF Staff Country Report No. 99/124, IMF, Washington, D.C.

International Institute for Population Sciences (1995), *National Family Health Survey, India 1992-93*. Demographic and Health Surveys. Bombay. The data are posted at http://www.measuredhs.com.

Klasen, Stephan (1999), *Malnourished and surviving in Asia, better nourished and dying young in Africa: What can explain this puzzle?* Revised version of Paper presented at the European Society for Population Economics Annual Meeting, June 24-27, 1999, Turin, Italy.

Knack, Stephen (1996), Institutions and the convergence hypothesis: the cross-national evidence. *Public choice* 87, No. 3, 207-228.

Knack, Stephen (1999), *IRIS-3 database*. PRS (Political Risk Services) Group, East Syracuse, N.Y. Available for sale at http://www.countrydata.com/datasets.

Knack, Stephen; Keefer, Philip (1995), Institutions and economic performance: cross-country tests using alternative institutional measures. *Economics & Politics* 7, No. 3, 207-227.

Kourgueni, I. A., Garba, B., Barrère, B. (1992), *Enquête Démographique et de Santé, Niger*. Demographic and Health Surveys. Niamey, Niger.

Latham, Michael C. (1997), *Human Nutrition in the Developing World*. Food and Nutrition Series No. 29. FAO, Rome.

Lind, Niels C. (1992), Some Thoughts on the Human Development Index. *Social Indicators Research* 27, 89-101.

Lüchters, Guido; Menkhoff, Lukas (1994), The Fourth Première of the Human Development Index. *Intereconomics* 30, No. 1, S. 9-15.

Lüchters, Guido; Menkhoff, Lukas (1996), Human Development as Statistical Artefact. *World Development* 24, No. 8, 1385-1392.

Lüchters, Guido; Menkhoff, Lukas (2000a), Chaotic signals from HDI measurement. *Applied Economics Letters* 7, 267-270.

Lüchters, Guido; Menkhoff, Lukas (2000b), The implicit equidistributional bias of human development. *Journal of international development* 12, No. 5, 613-623.

McGillivray, Mark; White, Howard (1993), Measuring Development? The UNDP's Human Development Index. *Journal of International Development* 5, No. 2, 183-192.

Messer, Ellen; Cohen, Marc J.; Marchione, Thomas (2001), Conflict: A Cause and Effect of Hunger. *Field Exchange* 12 (online journal). Posted at http://www.ennonline.net/fex.

MI/UNICEF/Tulane University (1998), *Progress in Controlling Vitamin A Deficiency*. Ottawa.

Ministry of Health and Family Welfare, Government of the People's Republic of Bangladesh (in Collaboration with Bangladesh National Nutrition Council) (1997), *Bangladesh Country Paper on Nutrition*. Bangladesh National Nutrition Council, Dhaka.

Mollica, Richard F. (2000), Invisible Wounds: Waging a New Kind of War. *Scientific American* 282, No. 6, 54-57.

Morris, Morris D. (1979), *Measuring the Condition of the World's Poor: The Physical Quality of Life Index*. Pergamon, New York.

Musau, Stephen N. (1999), *Community-Based Health Insurance: Experience and Lessons Learned from East Africa.* Technical Report No. 34. Partnerships for Health Reform Project, Abt Associates Inc., Bethesda, MD.

National Bureau of Statistics of China (1998), *China Population Statistics Yearbook 1998.* Beijing.

National Bureau of Statistics of China (1999), *China Statistical Yearbook 1999.* Beijing. Posted at http://www.stats.gov.cn/yearbook/ml/1999_e.htm.

Noorbakhsh, Farhad (1998), The Human Development Index: Some Technical Issues and Alternative Indices. *Journal of International Development* 10, 589-605.

Nubé, M. (2001), Confronting Dietary Energy Supply with Anthropometry in the Assessment of Undernutrition Prevalence at the Level of Countries. *World Development* 29, No. 7, 1275-1289.

Nübler, Irmgard (1995), The Human Development Index revisited. *Intereconomics*, July/August 1995, 173-176.

OFDA/CRED (2000), *International Database on Disasters*, posted at http://www.cred.be.

Osmani, Siddiq (1997), *Poverty and Nutrition in South Asia.* In: *Nutrition and Poverty, Papers from the ACC/SCN 24th Session Symposium.* Kathmandu.

Pelletier, David L.; Deneke, Kassahun; Kidane, Yemane; Haile, Beyenne; Negussie, Fikre (1995), The food-first bias and nutrition policy: lessons from Ethiopia. *Food Policy* 20, No. 4, 279-298.

Pelletier, David L.; Frongillo, Edward A., Jr.; Schroeder, Dirk G.; Habicht, Jean-Pierre (1994), A Methodology for Estimating the Contribution of Malnutrition to Child Mortality in Developing Countries. *The Journal of Nutrition* 124 (Supplement October), 2106 S – 2122 S.

Pinstrup-Andersen, Per; Pelletier, David; Alderman, Harold (eds.) (1995), *Child Growth and Nutrition in Developing Countries. Priorities for Action.* Cornell University Press, Ithaca.

Pritchett, Lant H.; Summers, Lawrence H. (1996), Wealthier is healthier. *Journal of Human Resources* 31, No. 4, 841-868.

Qizilbash, Mozaffar (1997), Pluralism and Well-being Indices. *World Development* 25, No. 12, 2009-2026.

Quisumbing, Agnes R.; Brown, Lynn R.; Feldstein, Hilary Sims; Haddad, Lawrence; Peña, Christine (1995), *Women: The Key to Food Security.* Food Policy Report. International Food Policy Research Institute (IFPRI), Washington, D.C.

Ramalingaswami, Vulimiri; Jonsson, Urban; Rohde, Jon (1996), *The Asian Enigma.* In: *UNICEF: Progress of Nations*. UNICEF, New York.

Rao, V. V. Bhanoji (1991), Human Development Report 1990: Review and Assessment. *World Development* 19, No. 10, 1451-1460.

Ravallion, Martin (1997), Good and Bad Growth: The Human Development Reports. *World Development* 25, No. 5, 631-638.

Reichel, Richard (1991), Der „Human Development Index" - ein sinnvoller Entwicklungsindikator? *Zeitschrift für Wirtschaftspolitik* 40, No. 1, 57-67.

Rizo, Alberto (1996), *Maternal mortality, induced abortion and sexually transmitted diseases: impact on women's health during the reproductive ages*. In: *United Nations, Population and Women*. Proceedings of the United Nations Expert Group Meeting on Population and Women, Gaborone, Botswana, 22-26 June 1992.

Rosegrant, Mark W.; Agcaoili-Sombilla, Mercedita; Perez, Nicostrato D. (1995), *Global Food Projections to 2020: Implications for Investment.* 2020 Vision Food, Agriculture and the Environment Discussion Paper No. 5. IFPRI, Washington, D.C.

Rosegrant, Mark W.; Paisner, Michael S., Meijer, Siet, Witcover, Julie (2001), *Global Food Projections to 2020: Emerging Trends and Alternative Futures*. IFPRI, Washington, D.C.

Ruel, Marie T.; Levin, Carol E.; Armar-Klemesu, Margaret; Maxwell, Daniel (1999), Good care practices can mitigate the negative effects of poverty and low maternal schooling on children's nutritional status: evidence from Accra. *World Development* 27, No. 11, 1993-2009.

Rwelamira, J. K.; Kleynhans, T. E. (1996), *SADC agricultural potential assessment: Country profiles*. Development Paper 124, Development Bank of South Africa, University of Stellenbosch, Maitland.

Ryten, Jacob (2000), *Should there be a Human Development Index?* Paper presented at the IAOS 2000 conference on 'Statistics, Human Rights and Development', Montreux, September 4-8, 2000.

Said, Ali (2000), *Monetary and non-monetary measures of poverty in Indonesia: A statistical comparison.* Paper presented at the IAOS 2000 conference on 'Statistics, Human Rights and Development', Montreux, September 4-8, 2000.

Santow, Gigi (1996), *Gender differences in health risks and use of services*. In: *United Nations, Population and Women.* Proceedings of the United Nations Expert Group Meeting on Population and Women, Gaborone, Botswana, 22-26 June 1992. UN, New York.

Schug, Walter (1991), *Äthiopien im Teufelskreis von Bevölkerungswachstum, Hunger und Umweltzerstörung.* In: Funkkolleg Humanökologie, Einführungsbrief. Tübingen.

Sen, Amartya K. (1976), Poverty: An Ordinal Approach to Measurement. *Econometrica* 44, 219-231.

Sen, Amartya K. (1990), Gender and Co-operative Conflicts. In: Irene Tinker (ed.), *Persistent Inequalities - Women and World Development*. New York, Oxford.

Sen, Amartya K. (1992), Missing Women - Social inequality outweighs women's survival advantage in Asia and North Africa. *British Medical Journal* 304, 587-588.

Sheng, Mingzhi (2002), *Lebensmittelkonsum und -konsumtrends in China – Eine empirische Analyse auf der Basis ökonometrischer Nachfragemodelle.* Peter Lang, Köln.

Shori, Rekha (2000), *Impact of socio-economic indicators on the health of children in Kenyan households with emphasis on poverty, malnutrition, childhood mortality, morbidity and HIV/AIDS.* Paper presented at the IAOS 2000 conference on 'Statistics, Human Rights and Development', Montreux, September 4-8, 2000.

Smith, Lisa C. (1998), Can FAO's measure of chronic undernourishment be strengthened? *Food policy* 23, No. 5, 425-445.

Smith, Lisa C.; Haddad, Lawrence (2000), *Explaining Child Malnutrition in Developing Countries: a Cross-Country Analysis*. IFPRI Research Report No. 111. IFPRI, Washington, D. C..

Spittler, Andreas; Holzer, Susanne; Roth, Erich (1995), *Ernährung und Immunologie (Immunonutrition).* In: Biesalski, Hans-Konrad; Fürst, Peter; Kasper, Heinrich; Kluthe, Reinhold; Pölert, Wolfgang; Puchstein, Christian; Stähelin, Hannes B. (eds.), *Ernährungsmedizin.* Thieme Verlag, Stuttgart.

Srinivasan, K. (2000), *Population: The Rising Graph.* In: Kejariwal, O. P. (ed.), *Indian Horizons.* Indian Council for Cultural Relations. Posted at http://www.hsph.harvard.edu/organizations/healthnet/SAsia/suchana/1028/rh328.htm.

Srinivasan, T.N (1994), Human Development: A New Paradigm or Reinvention of the Wheel? *AEA Papers and Proceedings* 84, No. 2, 238-243.

Streeten, Paul (1994), Human Development: Means and Ends. *AEA Papers and Proceedings* 84, No. 2, 232-237.

Subbarao, Kalinidhi; Raney, Laura (1995), Social gains from female education: A cross-national study. *Economic Development and Cultural Change* 44, No. 1, 105-128.

Svedberg, Peter (1998), *841 Million Undernourished? On the Tyranny of Deriving a Number*. Institute for International Economic Studies, Seminar Paper No. 656, Stockholm.

Szilágyi, György (2000), *International comparison theory – human development.* Paper presented at the IAOS 2000 conference on 'Statistics, Human Rights and Development', Montreux, September 4-8, 2000.

Tomkins, Andrew; Watson, Fiona (1989), *Malnutrition and Infection – a Review.* UN ACC/SCN, Geneva.

Trabold-Nübler, Harald (1991), The Human Development Index – A New Development Indicator? *Intereconomics*, September/October 1991, 236-243.

UN ACC/SCN (1993), *Second Report on the World Nutrition Situation: Country Trends, Methods and Statistics.* Volume II, Geneva.

UN ACC/SCN (1997), *Third Report on the World Nutrition Situation.* Geneva.

UN ACC/SCN (2000), *Fourth Report on the World Nutrition Situation: Nutrition Throughout the Life Cycle.* Geneva.

UNAIDS/WHO (1998), *Report on the Global HIV/AIDS epidemic – June 1998.* UNAIDS/WHO, New York, Geneva.

UNDP (1990), *Human Development Report 1990.* New York.

UNDP (1994), *Human Development Report 1994*. New York.

UNDP (1997), *Human Development Report 1997*. New York.

UNDP (1999), *Human Development Report 1999*. New York.

UNDP (2000), *Human Development Report 2000*. New York.

UNICEF (1990a), *World Declaration on the Survival, Protection and Development of Children and Plan of Action for Implementing the World Declaration on the Survival, Protection & Development of Children in the 1990s*. UNICEF, New York.

UNICEF (1990b), *Strategy for Improved Nutrition of Children and Women in Developing Countries*. UNICEF, New York.

UNICEF (1995), *The State of the World's Children 1995*. UNICEF, New York.

UNICEF (1998), *The State of the World's Children 1998*. UNICEF, New York.

UNICEF (1999), *The State of the World's Children 1999*. UNICEF, New York.

UNICEF (2001), *The State of the World's Children 2002*. UNICEF, New York.

UNICEF/UNU/WHO/MI (1999), *Preventing Iron Deficiency in Women and Children: Technical Consensus on Key Issues*. Technical Workshop, October 7-9, 1998. International Nutrition Foundation and MI, Boston and Ottawa.

von Braun, Joachim (1996), Den Hunger beenden und Natur erhalten – Vorschau auf den Welternährungsgipfel. *Spektrum der Wissenschaft*, No. 11, 76-82.

von Braun, Joachim; Bellin-Sesay, Friederike; Feldbrügge, Torsten; Heidhues, Franz (1998), *Verbesserung der Ernährung in Entwicklungsländern: Strategien und Politikempfehlungen*. Forschungsberichte des Bundesministeriums für wirtschaftliche Zusammenarbeit und Entwicklung, Band 123, Weltforum Verlag, Köln.

Waldron, Ingrid (1985), What do we know about causes of sex differences in mortality? A review of the literature. *Population Bulletin of the United Nations* 18, 59-76.

Waldron, Ingrid (1987), Patterns and causes of excess female mortality among children in developing countries. *World Health Statistics Quarterly* 40,194-210.

Webb, Patrick; von Braun, Joachim (1994), *Famine and Food Security in Ethiopia – Lessons for Africa*. John Wiley & Sons, Chichester.

WHO (1997), *WHO Global Database on Child Growth and Malnutrition, Programme of Nutrition*. WHO, Geneva.

WHO (1999), *World Health Report 1999: Making A Difference*. WHO, Geneva.

WHO (2000a), *WHO Global Database on Child Growth and Malnutrition*. Updates of this database are posted at http://www.who.int/nutgrowthdb/.

WHO (2000b), *HIV/AIDS database*. Posted at http://www.unaids.org/hivaidsinfo/statistics/june98/global_report/index.html

WHO (2001), *The Invisible Wounds: The Mental Health Crisis in Afghanistan*. WHO Special Report, Central Asia Crisis Unit, November 6. WHO, Geneva.

WHO/UNICEF/ICCIDD (1999), *Progress Towards the Elimination of Iodine Deficiency Disorders (IDD)*. WHO/NHD 99.4. WHO, Geneva.

Wiesmann, Doris; von Braun, Joachim; Feldbrügge, Torsten (2000), *An International Nutrition Index – Successes and Failures in Addressing Hunger and Malnutrition.* ZEF Discussion Paper on Development Policy No. 26. Center for Development Research, Bonn.

Wilson, Sara E. (2000), *AIDS Mushrooms into a Development Crisis.* 2020 Vision News and Views, December. IFPRI, Washington, D.C.

World Bank (1997), *Health, Nutrition and Population, Sector Strategy Studies.* Washington, D.C..

World Bank (1998), *World Development Indicators on CD-ROM*, Development Data Group, Washington, D.C.

World Bank (1999), *World Development Indicators on CD-ROM*, Development Data Group, Washington, D.C.

World Bank (2000), *World Development Indicators on CD-ROM*, Development Data Group, Washington, D.C.

Appendix

Box A 2.2-1 Formal proofs for the reduction of measurement error due to aggregation

The proofs of the assertions made in Box 2.2-1 are based on the following general mathematical and statistical rules:

(1) For two values A and B that are measured with errors ΔA and ΔB, respectively, $\Delta(A+B) = \Delta A + \Delta B$ (Fachredaktion des Bibliographischen Instituts 1985).

(2) If the expected value of variable Y is μ, the variance is σ^2 and the standard deviation is σ, then the linear transformation of Y, $Y^* = a + bY$, gives the corresponding parameters

$\mu^* = a + b\mu, \quad \sigma^{*2} = b^2\sigma^2, \quad \sigma^* = |b|\sigma$ (Bleymüller et al. 1994).

(3) The sum X of n independent normally distributed variables $X_1, \ldots, X_n$ is again normally distributed with

$\mu_X = \mu_1 + \ldots + \mu_n, \quad \sigma_X^2 = \sigma_1^2 + \ldots + \sigma_n^2$ (Bleymüller et al. 1994).

Consequently, $\sigma_X = \sqrt{\sigma_1^2 + \ldots + \sigma_n^2}$.

If an index is composed of three indicators I_1, I_2 and I_3 such as

$Index = \frac{1}{3}(I_1 + I_2 + I_3)$, then its measurement error is $\Delta Index = \frac{1}{3}(\Delta I_1 + \Delta I_2 + \Delta I_3)$.

As independent normal distributions are assumed for ΔI_1, ΔI_2 and ΔI_3, it follows that the mean value of $\Delta Index$:

$\mu_{\Delta \mathrm{Index}} = \frac{1}{3}(\mu_1 + \mu_2 + \mu_3)$,

with μ_1, μ_2 and μ_3 being the mean values of ΔI_1, ΔI_2 and ΔI_3, respectively. The standard deviation of $\Delta Index$ is derived from

$\sigma_{\Delta \mathrm{Index}} = \frac{1}{3}\sqrt{\sigma_1^2 + \sigma_2^2 + \sigma_3^2}$.

For equal standard deviations of ΔI_1, ΔI_2 and ΔI_3, that is, $\sigma_1 = \sigma_2 = \sigma_3$, we get $\sigma_{\Delta \mathrm{Index}} = \frac{1}{3}\sqrt{3\sigma_1^2} = \frac{\sqrt{3}}{3}\sigma_1 = \sqrt{\frac{1}{3}}\sigma_1$.

If $\sigma_1 = \sigma_2$, but $\sigma_3 > \sigma_1$, how many times larger than σ_1 may σ_3 be in order not to increase σ_{Index} above σ_1? The solution is obtained from

$\sigma_{\Delta \mathrm{Index}} = \frac{1}{3}\sqrt{2\sigma_1^2 + \sigma_3^2} \le \sigma_1$, which is equivalent to $2\sigma_1^2 + \sigma_3^2 \le 9\sigma_1^2 \Leftrightarrow \sigma_3 \le \sqrt{7}\sigma_1$.

Table A 4.1.2-1 Correction factors by country and region

Region, Subregion, Country	Year of Survey	Correction Factors			
		0-<3/ 0-<5 years	0.5-<3/ 0-<5 years	0.5-<5/0-<5 years	0-<4/0-<5 years
North Africa and Middle East					
Near East					
Bahrain	1989	0.92	..	..	1.01
Iraq	1991	1.06	1.23	1.09	1.05
Syria	1993	0.98	1.06	1.04	1.01
Turkey	1993	0.95	1.13	1.11	0.96
Yemen	1979	0.98	1.05	1.05	0.99
North Africa					
Algeria	1995	1.15	1.18	1.00	1.04
Egypt	1992-93	1.25	1.42	1.07	1.10
Egypt	1995-96	1.20	1.40	1.09	1.12
Egypt	1997-98	1.16	1.33	1.08	1.11
Morocco	1987	1.09	1.26	1.09	1.04
Morocco	1992	1.07	1.21	1.06	1.04
Tunisia	1973-75	1.04	1.14	1.05	1.05
Means*		*1.05*	*1.16*	*1.06*	*1.03*
Standard Deviations*		*0.08*	*0.09*	*0.03*	*0.04*
Sub-Saharan Africa					
West Africa					
Burkina Faso	1992-93	1.08	1.32	1.13	1.02
Côte d'Ivoire	1986	1.12	..	..	1.08
Ghana	1987-88	0.95	..	..	1.01
Mauritania	1988	0.98	1.07	1.06	1.02
Niger	1992	1.02	1.25	1.14	1.03
Nigeria	1990	0.98	1.18	1.12	1.00
Senegal	1992-93	0.97	1.20	1.14	1.00
Sierra Leone	1989	1.05	1.26	1.12	1.03
Sierra Leone	1990	1.03	1.22	1.12	1.02
East Africa					
Kenya	1993	1.03	1.17	1.08	1.02
Rwanda	1992	0.89	1.04	1.09	0.96
Tanzania	1991-92	1.00	1.17	1.11	1.01
Tanzania	1996	1.04	1.20	1.10	1.03
Uganda	1988-89	1.01	1.19	1.11	1.00
Central Africa					
Cameroon	1991	1.08	1.26	1.10	1.04
Congo, Rep.	1987	0.96	1.14	1.11	0.98
Southern Africa					
Lesotho	1992	0.91	1.06	1.11	0.96
Madagascar	1992	0.95	1.14	1.13	0.99
Malawi	1992	1.01	1.24	1.13	1.03
Namibia	1992	0.99	1.19	1.13	0.99
South Africa	1986	1.10	..	..	1.06
Zambia	1992	1.04	1.24	1.12	1.03
Zambia	1996-97	1.08	1.26	1.10	1.04
Means*		*1.00*	*1.18*	*1.11*	*1.01*
Standard Deviations*		*0.06*	*0.07*	*0.03*	*0.02*

Source Wiesmann et al. 2000

Note see next page (below Table A 4.1.2-1, continued)

Table A 4.1.2-1 (continued) Correction factors by country and region

Region, Subregion, Country	Year of Survey	Correction Factors 0-<3/ 0-<5 years	0.5-<3/ 0-<5 years	0.5-<5/0-<5 years	0-<4/0-<5 years
South, South-East and East Asia					
South Asia					
Bangladesh	1996-97	0.93	1,06	1.08	0.97
India	1974-79	0.91	..	..	0.97
India	1991-92	0.94	1.03	1.05	0.98
Pakistan	1990-91	0.92	1.05	1.09	0.97
South-East and East Asia					
Indonesia	1987	0.98	1.10	1.07	1.00
Indonesia	1995	0.92	1.04	1.07	0.99
Laos	1994	0.94	1.11	1.10	0.99
Mongolia	1997	0.84	1.06	1.17	1.04
Philippines	1989-90	0.98	1.10	1.07	0.98
Philippines	1993	0.93	1.10	1.11	1.00
Viet Nam	1983-84	0.95	..	..	0.99
Viet Nam	1998	0.89	0.99	1.07	0.96
Means*		*0.91*	*1.05*	*1.09*	*0.99*
Standard Deviations*		*0.04*	*0.04*	*0.04*	*0.03*
Latin America and Caribbean					
Central America and Caribbean					
Guatemala	1995	1.02	1.19	1.10	1.02
Haiti	1994-95	0.93	1.08	1.10	0.96
Honduras	1987	1.02	..	..	1.00
Honduras	1991-92	1.00	..	..	1.01
Mexico	1988	0.95	1.00	1.04	0.97
Nicaragua	1993	1.04	1.20	1.08	0.99
South America					
Colombia	1965-66	0.94	1.13	1.11	1.02
Colombia	1977-80	0.92	1.08	1.10	0.97
Colombia	1989	1.03	1.23	1.11	0.94
Colombia	1995	1.01	1.15	1.08	1.03
Guyana	1981	0.97	..	..	0.98
Guyana	1993	1.04	1.14	1.05	0.96
Peru	1991-92	1.05	1.23	1.09	0.98
Peru	1996	1.11	1.27	1.08	1.05
Means*		*1.01*	*1.15*	*1.08*	*1.00*
Standard Deviations*		*0.06*	*0.09*	*0.02*	*0.03*
All Regions					
Means*		*1.00*	*1.15*	*1.09*	*1.01*
Standard Deviations*		*0.07*	*0.09*	*0.03*	*0.03*

Source Wiesmann et al. 2000

Note * Means and standard deviations for regions were calculated from the most recent correction factors of each listed country (including only countries for which all four correction factors could be computed)

Table A 4.1.3-1-1 Dietary energy content of food categories for the Chinese diet

Food category	Dietary energy content (in kcal/kg)
Aquatic Products	530
Edible Oil	9000
Eggs and Related Products	1480
Grain	3170
Liquor	1250
Pork, Beef and Mutton	3330
Poultry	1710
Sugar	4000
Vegetables	230

Source author's own calculations based on data from Food Balance Sheet for China 1998 (FAO 2000b), and on data from Elmadfa et al. 1988

Figure A 4.2-1 Distribution of surveys and estimates in NI reference periods

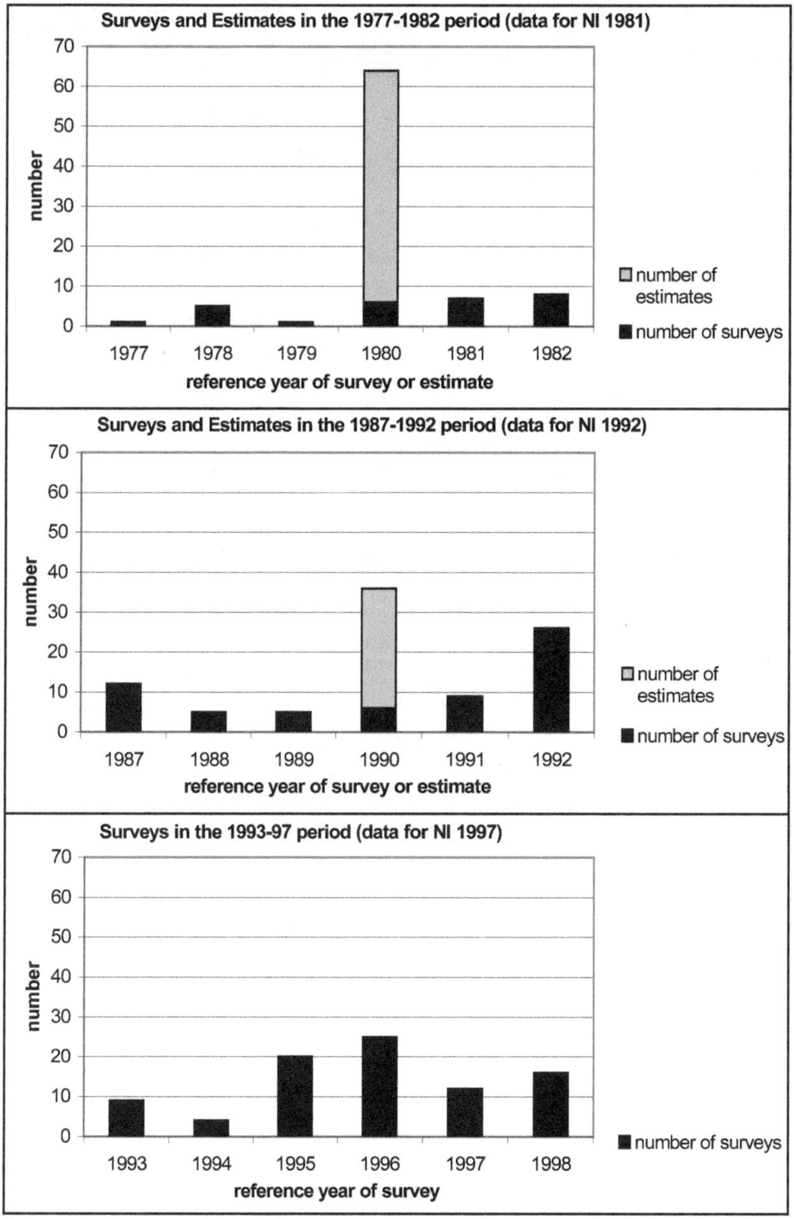

Source own presentation based on data from WHO 2000a and UN ACC/SCN 1993

Table A 5.1-1 NI for 1981, 1992, 1997 and underlying data

	Undernourished[1]			Underweight[2]			Under-5 mortality[3]			Nutrition Index		
	79-81	90-92	95-97	77-82	87-92	93-98	80	92	97	1981	1992	1997
Afghanistan	33.0	57.0	62.0	20.9 *	40.3 *	49.3	280	257	257	72.6	59.3	54.8
Albania	9.2 **	13.6 **	4.1 **	..	..	..	57	41	40	..	..	..
Algeria	9.0	5.0	5.0	18.0 *	9.2	12.8	145	68	39	86.2	93.0	92.9
Angola	29.0	50.0	43.0	26.3 *	35.3	36.3	261	292	292	72.9	62.0	64.0
Argentina	1.0	2.0	1.0	3.5 *	1.2	5.4	41	27	24	97.1	98.0	97.1
Armenia	..	..	24.8 **	..	..	3.3	..	33	30	..	..	89.7
Australia	..	..	..	..	..	..	13	8	6	..	..	..
Austria	..	..	..	..	..	..	17	8	5	..	..	..
Azerbaijan	..	..	27.6 **	..	..	9.3	..	52	45	..	..	86.4
Bahrain	..	..	..	..	..	..	..	..	22	..	..	..
Bangladesh	42.0	34.0	37.0	70.1	61.8	57.3	211	122	109	56.3	64.7	65.6
Belarus	..	..	..	..	..	..	..	..	18	..	..	..
Belgium	..	..	..	..	..	..	15	10	7	..	..	..
Benin	36.0	21.0	15.0	33.4 *	23.5 *	29.2	176	144	167	71.3	80.5	79.9
Bhutan	..	..	..	..	..	..	249	197	121	..	..	..
Bolivia	26.0	25.0	23.0	13.2	12.0	7.6	170	114	96	81.3	83.9	86.6
Bosnia & Herzeg.	..	..	21.0 **	..	..	..	..	..	16	..	..	..
Botswana	28.0	19.0	25.0	34.4 *	26.8 *	17.2	94	56	49	76.5	83.2	84.5
Brazil	15.0	13.0	10.0	7.0 *	7.0	5.7	93	63	44	89.6	91.3	93.3
Bulgaria	..	..	7.7 **	..	..	..	25	19	19	..	..	..
Burkina Faso	64.0	32.0	30.0	32.2 *	27.1 *	32.7	246	175	169	60.0	74.6	73.7
Burundi	38.0	44.0	63.0	25.9 *	31.0	..	193	178	176	72.4	69.3	..
Cambodia	62.0	41.0	33.0	44.3 *	37.7 *	47.4	330	181	167	53.8	68.1	68.1
Cameroon	20.0	30.0	32.0	17.3	15.1	20.6	173	113	99	81.8	81.3	79.4
Canada	..	..	..	..	..	..	13	8	7	..	..	..
Central Afric. R.	22.0	45.0	42.0	52.7 *	31.9 *	23.2	202	177	173	68.8	68.7	72.7
Chad	69.0	58.0	46.0	32.1 *	30.6 *	38.8	254	206	198	58.1	63.9	65.5
Chile	7.0	8.0	5.0	1.1	2.0 *	0.8	35	17	13	96.1	96.1	97.6
China	30.0	17.0	13.0	23.8 *	17.4	9.0	65	43	47	80.2	87.3	91.2
Colombia	12.0	17.0	12.0	16.7	10.1	8.4	59	19	30	88.6	90.5	92.3
Comoros	56.6 **	51.4 **	51.8 **	..	18.5	25.8	..	182 x	93	..	70.7	71.4
Congo, Dem. R.	37.0	36.0	55.0	27.9 *	33.2 *	34.4	204	187	207	71.7	71.0	63.6
Congo, Rep.	29.0	32.0	34.0	39.1 *	23.9	..	125	109	108	73.5	78.0	..
Costa Rica	8.0	6.0	7.0	6.0	2.3	4.1	29	16	14	94.4	96.7	95.9
Côte d'Ivoire	7.0	14.0	15.0	14.1 *	12.3 *	21.3	180	120	150	86.9	87.2	83.0
Croatia	..	..	16.6 **	..	..	0.6	..	..	9	..	..	94.0
Cuba	3.0	3.0	19.0	8.3 *	8.4 *	..	26	10	8	95.4	96.0	..
Cyprus	..	..	..	..	..	..	..	10 x	9	..	..	..
Czech Republic	..	..	..	..	..	..	..	10	7	..	..	..
Denmark	..	..	..	..	..	..	10	7	6	..	..	..
Djibouti	55.4 **	57.3 **	34.4 **	..	22.9	..	..	158 x	156	..	68.2	..
Dominican Rep.	25.0	28.0	26.0	14.0 *	10.3	5.9	94	48	53	84.0	85.8	87.7
Ecuador	12.0	8.0	5.0	..	16.5	..	101	57	39	..	90.1	..

Source FAO 1999a, WHO 2000a, UN ACC/SCN 1993, UNICEF 1995 and 1999; author's own calculations based on FAO 1999a and 1999b

Note 1 in the population (in %)
2 in children under five (in %)
3 per 1000 live births
* estimates reported in UN ACC/SCN 1993
** own estimates
x data refer to 1993

Table A 5.1-1 (continued) NI for 1981, 1992, 1997 and underlying data

	Undernourished[1]			Underweight[2]			Under-5 mortality[3]			Nutrition Index		
	79-81	90-92	95-97	77-82	87-92	93-98	80	92	97	1981	1992	1997
Egypt	8.0	4.0	4.0	14.9	10.4	10.7	180	59	73	86.3	93.3	92.7
El Salvador	17.0	12.0	10.0	20.9 *	15.2	11.8	120	60	36	83.5	89.1	91.7
Eritrea	..	..	67.0	..	..	43.7	260	204	116	..	..	59.8
Estonia	..	..	6.6 **	..	..	..	..	23	14	..	..	..
Ethiopia	53.5 **	67.9 **	51.0	38.1	43.8	..	260	204	175	61.0	56.5	..
Fiji	..	10.4 **	6.1 **	..	..	7.9	..	..	24	..	..	94.6
Finland	..	..	..	..	..	..	9	5	4	..	..	..
France	..	..	..	..	..	..	13	9	5	..	..	..
Gabon	13.0	11.0	8.0	16.1 *	15.1 *	..	194	154	145	83.8	86.1	..
Gambia	57.0	17.0	25.0	25.6 *	17.1 *	26.2	..	216 x	87	..	81.4	80.3
Georgia	..	..	12.8 **	..	..	..	..	28	29	..	..	..
Germany	..	..	..	..	..	..	16	7	5	..	..	..
Ghana	61.0	29.0	11.0	30.9 *	27.1	27.3	157	170	107	64.5	75.8	83.9
Greece	..	..	..	..	..	..	23	10	8	..	..	..
Guatemala	17.0	14.0	17.0	43.6	28.5	26.6	136	73	55	75.7	83.7	84.0
Guinea	30.0	37.0	31.0	23.4	24.0 *	..	276	226	201	73.0	72.2	..
Guinea-Bissau	39.6 **	15.9 **	17.3 **	..	..	..	290	235	220	..	..	..
Guyana	13.0	24.0	16.0	22.1	18.0 *	18.3	..	63 x	82	..	84.1	86.0
Haiti	47.0	63.0	61.0	37.4	26.8	27.5	195	130	132	65.7	66.1	66.5
Honduras	31.0	23.0	21.0	21.2 *	18.0	25.4	100	56	45	79.5	84.7	83.4
Hungary	..	..	..	..	..	..	26	15	11	..	..	..
India	38.0	26.0	22.0	68.0 *	61.0	45.4	177	122	108	59.5	67.6	74.4
Indonesia	26.0	10.0	6.0	45.7 *	39.9	34.0	128	111	68	72.3	80.0	84.8
Iran	9.0	6.0	6.0	43.1	..	10.9	126	54	35	78.8	..	93.3
Iraq	4.0	9.0	15.0	14.5 *	11.9	..	83	71	122	91.1	90.7	..
Ireland	..	..	..	..	..	..	14	7	7	..	..	..
Israel	..	..	..	..	..	..	19	9	6	..	..	..
Italy	..	..	..	..	..	..	17	9	6	..	..	..
Jamaica	8.0	12.0	11.0	9.3	4.6	4.2	39	13	11	93.0	94.1	94.6
Japan	..	..	..	..	..	..	11	6	6	..	..	..
Jordan	6.0	4.0	3.0	..	6.4	5.1	66	27	24	..	95.7	96.5
Kazakhstan	..	..	2.8 **	..	..	8.3	..	49	44	..	..	94.9
Kenya	25.0	47.0	41.0	22.0	13.2	22.1	112	90	87	80.8	77.1	76.4
Korea, Dem. R.	19.0	16.0	48.0	..	..	..	43	32	30	..	..	..
Korea, Rep.	1.0	1.0	1.0	..	..	..	18	9	6	..	..	..
Kuwait	4.0	27.0	3.0	10.1 *	5.0 *	1.7	35	13	13	94.2	89.0	98.0
Kyrgyz Rep.	..	..	23.4 **	..	..	11.0	..	58	48	..	..	87.1
Laos	32.0	31.0	33.0	37.6 *	34.0 *	40.0	190	141	122	70.8	74.0	72.0
Latvia	..	..	5.0 **	..	..	..	..	26	20	..	..	..
Lebanon	8.0	2.0	2.0	14.0 *	8.9 *	3.0	40	40	37	91.5	95.1	97.1
Lesotho	26.0	31.0	28.0	13.3	15.8	16.0	173	156	137	81.1	79.3	80.8
Liberia	22.0	49.0	42.0	20.8 *	20.1 *	..	235	217	235	77.9	69.8	..
Libya	0.0	1.0	1.0	4.1 *	4.0 *	4.7	150	100	25	93.4	94.9	97.3

Source FAO 1999a, WHO 2000a, UN ACC/SCN 1993, UNICEF 1995 and 1999; author's own calculations based on FAO 1999a and 1999b

Note 1 in the population (in %)
2 in children under five (in %)
3 per 1000 live births

* estimates reported in UN ACC/SCN 1993

** own estimates

x data refer to 1993

Table A 5.1-1 (continued) NI for 1981, 1992, 1997 and underlying data

	Undernourished[1]			Underweight[2]			Under-5 mortality[3]			Nutrition Index		
	79-81	90-92	95-97	77-82	87-92	93-98	80	92	97	1981	1992	1997
Lithuania	..	..	..	..	..	..	..	20	15	..	..	..
Macedonia	..	..	11.3 **	..	..	..	..	..	23	..	..	..
Madagascar	18.0	34.0	39.0	30.1 *	40.9	40.0	216	164	158	76.9	70.0	68.8
Malawi	26.0	45.0	37.0	21.2	27.6	29.9	290	223	215	74.5	68.5	70.7
Malaysia	4.0	3.0	2.0	29.8 *	25.6	20.1	42	17	11	87.7	90.2	92.5
Mali	59.0	30.0	29.0	34.3 *	25.1	40.0	310	217	239	58.7	74.5	69.3
Mauritania	35.0	15.0	13.0	31.0	47.6	23.0	249	202	183	69.8	72.7	81.9
Mauritius	10.0	6.0	6.0	28.0 *	17.0 *	14.9	42	22	23	86.3	91.8	92.4
Mexico	5.0	5.0	6.0	16.7 *	14.2	..	81	32	35	90.2	92.7	..
Moldova	..	..	13.7 **	..	..	..	..	36	31	..	..	..
Mongolia	27.0	34.0	48.0	..	12.3	..	112	78	150	..	82.1	..
Morocco	10.0	5.0	5.0	16.6 *	9.5	..	145	59	72	86.3	93.2	..
Mozambique	54.0	66.0	63.0	43.8 *	46.8 *	26.1	269	282	208	58.7	53.4	63.6
Myanmar	19.0	9.0	7.0	42.0	32.4	28.2	146	111	114	75.2	82.8	84.7
Namibia	25.0	26.0	30.0	..	26.2	..	114	79	75	..	80.3	..
Nepal	46.0	21.0	21.0	66.2 *	50.5 *	46.9	177	128	104	57.4	72.4	74.4
Netherlands	..	..	..	..	..	..	11	8	6	..	..	..
New Zealand	..	..	..	..	..	..	16	9	7	..	..	..
Nicaragua	26.0	29.0	31.0	10.0	..	12.2	143	72	57	83.2	..	83.9
Niger	32.0	41.0	39.0	49.0 *	42.6	49.6	320	320	320	62.6	61.6	60.1
Nigeria	40.0	13.0	8.0	30.4 *	35.3	35.0	196	191	187	70.2	77.7	79.6
Norway	..	..	..	..	..	..	11	8	4	..	..	..
Oman	..	..	..	..	24.3	..	95	29	18	..	..	..
Pakistan	31.0	20.0	19.0	50.2	40.2	38.2	151	137	136	68.4	75.7	76.7
Panama	22.0	18.0	17.0	15.7	6.1	..	31	20	20	86.6	91.4	..
Papua New Guin.	31.0	27.0	24.0	29.9	..	..	95	95	112	76.9	..	..
Paraguay	13.0	18.0	13.0	7.0 *	3.7	..	61	34	33	91.3	91.7	..
Peru	28.0	40.0	19.0	16.7 *	10.7	7.8	130	62	56	80.9	81.2	89.3
Philippines	27.0	21.0	22.0	33.2	33.4	29.6	70	59	41	78.0	80.3	81.8
Poland	..	..	..	..	..	..	24	15	11	..	..	..
Portugal	..	..	..	..	..	..	31	11	8	..	..	..
Qatar	..	..	..	..	..	..	..	..	20	..	..	..
Romania	..	3.3 **	..	..	5.7	..	36	29	26	..	96.1	..
Russian Federat.	..	..	5.2 **	..	..	3.0	..	31	25	..	..	96.4
Rwanda	24.0	40.0	37.0	..	29.4	27.3	222	141	170	..	72.5	73.1
Saudi Arabia	3.0	3.0	4.0	14.9 *	12.6 *	..	90	38	28	91.1	93.6	..
Senegal	19.0	19.0	17.0	19.4 *	21.6	22.3	221	120	124	79.8	82.6	82.9
Sierra Leone	40.0	44.0	43.0	23.2	28.7	..	301	284	316	68.8	66.4	..
Singapore	..	..	..	..	..	..	13	6	4	..	..	..
Slovak Rep.	..	..	4.2 **	..	..	..	..	18	11	..	..	..
Slovenia	..	..	..	..	..	..	..	..	6	..	..	..
Somalia	55.0	70.0	73.0	41.9 *	38.8 *	..	246	211	211	59.8	57.1	..
South Africa	5.5	4.9	4.1 **	..	..	8.2	91	69	65	..	..	93.7

Source FAO 1999a, WHO 2000a, UN ACC/SCN 1993, UNICEF 1995 and 1999; author's own calculations based on FAO 1999a and 1999b

Note 1 in the population (in %)
2 in children under five (in %)
3 per 1000 live births
* estimates reported in UN ACC/SCN 1993
** own estimates

Table A 5.1-1 (continued) NI for 1981, 1992, 1997 and underlying data

	Undernourished[1]			Underweight[2]			Under-5 mortality[3]			Nutrition Index		
	79-81	90-92	95-97	77-82	87-92	93-98	80	92	97	1981	1992	1997
Spain	..	..	..	..	..	..	16	9	5	..	..	..
Sri Lanka	22.0	29.0	25.0	47.5	37.3	37.7	52	19	19	75.7	77.8	79.0
Sudan	24.0	31.0	20.0	26.4 *	33.7 *	33.9	200	128	115	76.6	74.5	78.5
Swaziland	14.0	9.0	14.0	12.6 *	8.8 *	..	..	107 x	94	..	90.5	..
Sweden	..	..	..	..	..	..	9	6	4	..	..	..
Switzerland	..	..	..	..	..	..	11	8	5	..	..	..
Syria	3.0	2.0	1.0	16.0 *	12.5 *	12.9	73	39	33	91.3	94.0	94.4
Tajikistan	..	..	32.5 **	..	..	..	..	83	76	..	..	..
Tanzania	23.0	30.0	40.0	23.8 *	28.9	30.6	202	167	143	77.7	75.0	72.0
Thailand	28.0	27.0	24.0	36.0	22.2	18.6	61	33	38	77.1	82.8	84.8
Togo	31.0	29.0	23.0	23.2 *	24.6	22.2	175	135	125	76.2	77.8	80.9
Trinidad & Tobago	5.0	12.0	11.0	10.0 *	6.7	..	40	21	17	93.8	93.2	..
Tunisia	2.0	1.0	1.0	14.8 *	10.3	9.0	102	36	33	91.0	95.1	95.6
Turkey	2.0	2.0	2.0	13.2 *	10.5 *	8.3	141	84	45	90.2	93.0	95.1
Turkmenistan	..	..	17.5 **	..	..	..	..	89	78	..	..	..
Uganda	31.0	23.0	28.0	24.8 *	23.0	25.5	181	185	137	75.5	78.6	77.8
Ukraine	..	..	6.6 **	..	..	..	..	25	24	..	..	..
United Arab Em.	1.0	2.0	1.0	..	..	..	64	21	10	..	..	..
United Kingdom	..	..	..	..	..	..	14	8	7	..	..	..
United States	..	..	..	..	..	..	15	10	8	..	..	..
Uruguay	3.0	7.0	4.0	6.5 *	7.4	4.4	42	21	21	95.5	94.6	96.5
Uzbekistan	..	..	14.5 **	..	..	18.8	..	66	60	..	..	87.1
Venezuela	4.0	11.0	15.0	10.2	5.1	8.1	42	24	25	93.9	93.9	91.6
Vietnam	33.0	28.0	19.0	53.1 *	41.9	39.8	105	48	43	68.4	75.7	79.5
Yemen	40.0	37.0	37.0	55.7	30.0	46.1	210	137	100	61.6	73.4	69.5
Yugoslavia	2.6 **	..	3.1 **	..	..	1.6	37	22	21	..	..	97.7
Zambia	30.0	39.0	45.0	19.3 *	25.2	23.5	160	203	202	78.3	72.0	70.6
Zimbabwe	30.0	40.0	39.0	23.5 *	12.0	15.5	125	83	80	78.2	80.1	79.4

Source FAO 1999a, WHO 2000a, UN ACC/SCN 1993, UNICEF 1995 and 1999; author's own calculations based on FAO 1999a and 1999b

Note 1 in the population (in %)
2 in children under five (in %)
3 per 1000 live births
* estimates reported in UN ACC/SCN 1993
** own estimates
x data refer to 1993

Table A 6.2.2-1 Spearman rank correlation coefficients for poverty measure ranking differences and ranking differences of the NI and other development indicators (vis-à-vis GNP per capita ranking, for NI 1997 only)

	NI DIFF	$NI_{2,3}$ DIFF	PUN DIFF	DES DIFF	HDI DIFF	HDI* DIFF
Ranking differences to GNP per capita						
Population below 1$ a day (%)	0.499 *** (0.001)	0.320 ** (0.039)	0.464 *** (0.002)	0.380 ** (0.013)	0.205 (0.193)	0.233 (0.137)
Poverty gap at 1$ a day (%)	0.443 *** (0.003)	0.243 (0.120)	0.476 *** (0.001)	0.422 *** (0.005)	0.197 (0.212)	0.239 (0.128)
Population below 2$ a day (%)	0.465 *** (0.002)	0.444 *** (0.003)	0.292 * (0.061)	0.201 (0.202)	0.131 (0.409)	0.158 (0.319)
Poverty gap at 2$ a day (%)	0.512 *** (0.001)	0.382 ** (0.012)	0.436 *** (0.004)	0.351 ** (0.023)	0.197 (0.210)	0.230 (0.143)
Number of cases: 42						

Source author's own calculation based on data described in Chapter 6.2

Table A 6.2.2-2 Spearman rank correlation coefficients for poverty measure ranking differences and NI ranking differences vis-à-vis HDI and HDI*, respectively (for NI 1997 only)

	NI-HDI DIFF		NI-HDI* DIFF
Ranking differences to HDI ranking		*Ranking differences to HDI* ranking*	
Population below 1$ a day (%)	0.424 *** (0.005)	**Population below 1$ a day (%)**	0.434 *** (0.004)
Poverty gap at 1$ a day (%)	0.380 ** (0.013)	**Poverty gap at 1$ a day (%)**	0.380 ** (0.013)
Population below 2$ a day (%)	0.527 *** (0.000)	**Population below 2$ a day (%)**	0.545 *** (0.000)
Poverty gap at 2$ a day (%)	0.493 *** (0.001)	**Poverty gap at 2$ a day (%)**	0.519 *** (0.000)
Number of cases: 42			

Source author's own calculation based on data described in Chapter 6.2

Table A 6.2.2-3 Spearman rank correlation coefficients for poverty measure ranking differences and ranking differences of the NI and other development indicators (for restricted data set with poverty data collection in 1992-93 or 1996-97)

	NI DIFF	$NI_{2,3}$ DIFF	PUN DIFF	DES DIFF	HDI DIFF	HDI* DIFF
Ranking differences to GNP per capita						
Population below 1$ a day (%)	0.558 *** (0.009)	0.584 *** (0.005)	0.480 ** (0.028)	0.452 ** (0.040)	0.205 (0.372)	0.273 (0.232)
Poverty gap at 1$ a day (%)	0.372 * (0.097)	0.412 * (0.063)	0.425 * (0.055)	0.487 ** (0.025)	0.056 (0.810)	0.141 (0.543)
Population below 2$ a day (%)	0.513 ** (0.017)	0.663 *** (0.001)	0.181 (0.432)	0.133 (0.565)	0.264 (0.274)	0.316 (0.163)
Poverty gap at 2$ a day (%)	0.488 ** (0.025)	0.550 *** (0.010)	0.343 (0.128)	0.339 (0.133)	0.184 (0.426)	0.247 (0.279)
Number of cases: 21						

Source author's own calculation based on data described in Chapter 6.2

Table A 6.2.2-4 Spearman rank correlation coefficients for poverty measure ranking differences and NI ranking differences vis-à-vis HDI and HDI*, respectively (for restricted data set with poverty data collection in 1992-93 or 1996-97)

	NI-HDI DIFF		NI-HDI* DIFF
Ranking differences to HDI ranking		*Ranking differences to HDI* ranking*	
Population below 1$ a day (%)	0.588 *** (0.005)	**Population below 1$ a day (%)**	0.482 ** (0.027)
Poverty gap at 1$ a day (%)	0.499 ** (0.021)	**Poverty gap at 1$ a day (%)**	0.415 * (0.061)
Population below 2$ a day (%)	0.593 *** (0.005)	**Population below 2$ a day (%)**	0.621 *** (0.003)
Poverty gap at 2$ a day (%)	0.547 *** (0.010)	**Poverty gap at 2$ a day (%)**	0.526 ** (0.014)
Number of cases: 21			

Source author's own calculation based on data described in Chapter 6.2

Box A 7.1.3-1 Vector and matrix notation of regression equations

A set of N linear equations for the regression of a dependent variable y with N observations on k independent variables can be written as follows:

$$
\begin{array}{ll}
n=1 & y_1=\beta_0+\beta_1 x_{11}+\beta_2 x_{21}+\ldots+\beta_k x_{k1}+u_1 \\
n=2 & y_2=\beta_0+\beta_1 x_{12}+\beta_2 x_{22}+\ldots+\beta_k x_{k2}+u_2 \\
\vdots & \vdots \\
n=N & y_N=\beta_0+\beta_1 x_{1N}+\beta_2 x_{2N}+\ldots+\beta_k x_{kN}+u_N
\end{array}
$$

An alternative notation uses matrix algebra:

$$
\begin{bmatrix} y_1 \\ y_2 \\ \vdots \\ y_N \end{bmatrix} = \begin{bmatrix} 1 & x_{11} & x_{21} & \cdots & x_{k1} \\ 1 & x_{12} & x_{22} & \cdots & x_{k2} \\ \vdots & \vdots & \vdots & & \vdots \\ 1 & x_{1N} & x_{2N} & \cdots & x_{kN} \end{bmatrix} \cdot \begin{bmatrix} \beta_0 \\ \beta_1 \\ \vdots \\ \beta_k \end{bmatrix} + \begin{bmatrix} u_1 \\ u_2 \\ \vdots \\ u_N \end{bmatrix}
$$

The short form is

$$y=\mathbf{X}\beta+u\,.$$

In this notation, y denominates the vector of N observations of the dependent variable, u the vector of error terms, **X** the data matrix of independent variables containing the observations of each single variable x_i in the i-th column of **X,** and β the vector of parameters (adapted from Hansen 1993).

Box A 7.2.2-1 Excerpts from variable descriptions, International Country Risk Guide

Corruption in Government

Lower scores indicate that 'high government officials are likely to demand special payments' and that 'illegal payments are generally expected throughout lower levels of government' in the form of 'bribes connected with import and export licenses, exchange controls, tax assessment, police protection, or loans.'

Rule of Law (named 'Law and Order Tradition' in ICRG)

This variable 'reflects the degree to which the citizens of a country are willing to accept the established institutions to make and implement laws and adjudicate disputes.' Higher scores indicate: 'sound political institutions, a strong court system, and provisions for an orderly succession of power.' Lower scores indicate: 'a tradition of depending on physical force or illegal means to settle claims.' Upon changes in government new leaders 'may be less likely to accept the obligations of the previous regime.'

Quality of the Bureaucracy

High scores indicate 'an established mechanism for recruitment and training,' 'autonomy from political pressure,' and 'strength and expertise to govern without drastic changes in policy or interruptions in government services' when governments change.

Risk of Repudiation of Contracts by Government

'This indicator addresses the possibility that foreign businesses, contractors, and consultants face the risk of a modification in a contract taking the form of a repudiation, postponement, or scaling down' due to 'an income drop, budget cutbacks, indigenisation pressure, a change in government, or a change in government economic and social priorities.' Lower scores signify 'a greater likelihood that a country will modify or repudiate a contract with a foreign business.'

Risk of Expropriation of Private Investment

This variable evaluates the risk of 'outright confiscation and forced nationalisation' of property. Lower ratings 'are given to countries where expropriation of private foreign investment is a likely event.'

Quoted from Knack (1999).

Table A 7.2.2-1 Classification of countries with NI 1997 or NI 1992 by region

Africa and Middle East			Asia and Europe			America	
Middle East & North Africa	**Sub-Saharan Africa**		**Eastern Europe & former Soviet Union**	**South Asia**	**East Asia & Pacific**	**The Caribbean**	**Latin America**
Algeria	Angola	Madagascar	Armenia	Afghanistan	Cambodia	Costa Rica	Argentina
Egypt	Benin	Malawi	Azerbaijan	Bangladesh	China	Cuba	Bolivia
Iran	Botswana	Mali	Croatia	India	Fiji	Dominican Rep.	Brazil
Iraq	Burkina Faso	Mauritania	Kazakhstan	Myanmar	Indonesia	El Salvador	Chile
Jordan	Burundi	Mauritius	Kyrgyz Republic	Nepal	Laos	Guatemala	Colombia
Kuwait	Cameroon	Mozambique	Romania	Pakistan	Malaysia	Haiti	Ecuador
Lebanon	Central African Rep.	Namibia	Russian Federation	Sri Lanka	Mongolia	Honduras	Guyana
Libya	Chad	Niger	Uzbekistan		Philippines	Jamaica	Paraguay
Morocco	Comoros	Nigeria	Yugoslavia		Thailand	Mexico	Peru
Saudi Arabia	Congo, Dem. Rep.	Rwanda			Vietnam	Nicaragua	Trinidad & Tobago
Syria	Congo, Rep.	Senegal				Panama	Uruguay
Tunisia	Côte d'Ivoire	Sierra Leone					Venezuela
Turkey	Djibouti	Somalia					
Yemen	Eritrea	South Africa					
	Ethiopia	Sudan					
	Gabon	Swaziland					
	The Gambia	Tanzania					
	Ghana	Togo					
	Guinea	Uganda					
	Kenya	Zambia					
	Lesotho	Zimbabwe					
	Liberia						

Source author's own presentation based on the classification of countries by the World Bank (World Bank 2000)

Table A 7.3.1-1 Correlation matrix of the variables in the initial model (Pearson's correlation coefficients), including NI 1997 supplemented by NI 1992, independent variables matched

	NI	logarithm of GNP per capita	urban population	immunisation against DPT	military expenditure	war dummy	political freedom index	Middle East & North Africa	South Asia	Eastern Europe & FSU
Nutrition Index	1	-	-	-	-	-	-	-	-	-
logarithm of GNP per capita (in constant 1995 PPP$)	0.862 *** (0.000)	1	-	-	-	-	-	-	-	-
urban population (in % of total population)	0.757 *** (0.000)	0.705 *** (0.000)	1	-	-	-	-	-	-	-
one-year olds fully immunised against DPT[1] (in %)	0.648 *** (0.000)	0.573 *** (0.000)	0.366 *** (0.000)	1	-	-	-	-	-	-
military expenditure (in % of GNP)	-0.078 (0.454)	0.039 (0.707)	0.081 (0.435)	-0.088 (0.398)	1	-	-	-	-	-
war dummy[2] (0 for non-war countries)	0.231 ** (0.025)	-0.171 * (0.099)	-0.077 (0.458)	-0.081 (0.440)	0.239 ** (0.020)	1	-	-	-	-
political freedom index	0.199 * (0.055)	-0.302 *** (0.003)	-0.239 ** (0.021)	-0.251 ** (0.015)	0.286 *** (0.005)	0.238 ** (0.021)	1	-	-	-
Middle East & North Africa dummy	0.318 *** (0.002)	0.215 ** (0.038)	0.282 *** (0.006)	0.203 ** (0.050)	0.360 *** (0.000)	0.090 (0.386)	0.247 ** (0.016)	1	-	-
South Asia dummy	0.198 * (0.056)	-0.089 (0.395)	0.247 ** (0.016)	0.131 (0.210)	0.006 (0.951)	-0.030 (0.774)	-0.161 (0.122)	-0.086 (0.408)	1	-
Eastern Europe & former Soviet Union dummy	0.245 ** (0.017)	0.143 (0.169)	0.155 (0.135)	0.260 ** (0.011)	0.023 (0.829)	0.171 * (0.099)	-0.096 (0.360)	-0.111 (0.287)	-0.072 (0.489)	1

Number of cases: 94

Notes 1 Diphtheria, Pertussis, Tetanus
2 refers to wars begun, ended or continued in 1983-92 or 1988-97 (matched with NI data)

Source author's own calculation based on data described in Chapter 7.2

Table A 7.3.1-2 Initial model with war dummy variable relating to a five-year reference period

Dependent variable				
Nutrition Index (NI) 1997 supplemented by NI 1992				
Independent variables	**Coefficients**			
	non-standard.	**standard.**	**T-statistics**	**Significance**
logarithm of GNP per capita in constant 1995 PPP$	6.59	0.52	8.50 ***	0.000
urban population (in % of total population)	0.14	0.26	4.67 ***	0.000
one-year olds fully immunised against DPT[1] (in %)	0.11	0.21	4.16 ***	0.000
military expenditure (in % of GNP)	-0.43	-0.14	-3.18 ***	0.002
war dummy[2] (0 for non-war countries)	-2.95	-0.11	-2.63 ***	0.010
political freedom index	0.48	0.08	1.81 *	0.074
Regional dummies				
Middle East and North Africa	4.09	0.12	2.70 ***	0.008
South Asia	-3.83	-0.08	-1.99 **	0.050
Eastern Europe and former Soviet Union	4.02	0.11	2.49 **	0.015
Intercept	16.67		3.25 ***	0.002
Model statistics				
Number of observations	94			
R-squared, adjusted	0.872			
Standard error	3.816			
F statistic	71.638			

Source author's own calculation based on data described in Chapter 7.2

Notes 1 Diphtheria, Pertussis, Tetanus
2 refers to wars begun, ended or continued in 1983-92 or 1988-97 (matched with NI data)

Table A 7.3.1-3 Model including war duration variable for a subset of war countries according to the five-year reference period

Dependent variable				
Nutrition Index (NI) 1997 supplemented by NI 1992				
Independent variables	**Coefficients**			
	non-standard.	**standard.**	**T-statistics**	**Significance**
logarithm of GNP per capita in constant 1995 PPP$	7.93	0.54	4.43 ***	0.001
urban population (in % of total population)	0.17	0.27	2.74 **	0.017
one-year olds fully immunised against DPT[1] (in %)	0.14	0.27	3.15 ***	0.008
duration of war[2] in years (0 for non-war countries)	-0.22	-0.14	-2.15 *	0.051
Intercept	3.59		0.37	0.721
Model statistics				
Number of observations	18			
R-squared, adjusted	0.939			
Standard error	3.270			
F statistic	66.456			

Source author's own calculation based on data described in Chapter 7.2

Notes 1 Diphtheria, Pertussis, Tetanus
2 refers to wars begun, ended or continued in 1983-92 or 1988-97 (matched with NI data)

Table A 7.3.2-1 Correlation matrix of residuals in regressions of NI components

		Residuals in Regressions on Basic Model Explanatory Variables for PUN	CUW	CM
Residuals in Regressions on Basic Model Explanatory Variables for	PUN	1		
	CUW	-0.120 (0.251)	1	
	CM	-0.092 (0.379)	0.140 (0.177)	1
Number of Cases: 94				

Source author's own calculation based on data described in Chapter 7.2

Table A 7.3.2-2 Actual and predicted values of coefficients in the basic NI model

	Non-standardised Coefficients in Basic NI Model: Actual	Predicted	Deviation in %
logarithm of GNP per capita in constant 1995 PPP$	6.73456	6.73490	0.01
urban population (in % of total population)	0.13515	0.13512	-0.02
one-year olds fully immunised against DPT[1] (in %)	0.11467	0.11471	0.03
military expenditure (in % of GNP)	-0.43166	-0.43186	0.05
duration of war[2] in years (0 for non-war countries)	-0.23163	-0.23162	-0.00
political freedom index	0.54003	0.53916	-0.16
Regional dummies			
Middle East and North Africa	3.65721	3.65761	0.01
South Asia	-4.20095	-4.19943	-0.04
Eastern Europe and former Soviet Union	2.62285	2.62352	0.03
Intercept	15.07426	15.07274	-0.01

Source author's own calculation based on data described in Chapter 7.2

Notes 1 Diphtheria, Pertussis, Tetanus
2 refers to wars begun, ended or continued in 1983-92 or 1988-97 (matched with NI data)

Table A 7.3.3-1 Model including the additional variable 'births attended by health staff'

Dependent variable				
Nutrition Index (NI) 1997 supplemented by NI 1992				
Independent variables	**Coefficients**			
	non-standard.	**standard.**	**T-statistics**	**Significance**
logarithm of GNP per capita in constant 1995 PPP$	5.62	0.44	7.19 ***	0.000
urban population (in % of total population)	0.13	0.25	4.48 ***	0.000
births attended by health staff (in %)	0.08	0.21	3.61 ***	0.001
one-year olds fully immunised against DPT[1] (in %)	0.08	0.16	3.38 ***	0.001
military expenditure (in % of GNP)	-0.51	-0.16	-3.93 ***	0.000
political freedom index	0.81	0.14	3.35 ***	0.001
duration of war[2] in years (0 for non-war countries)	-0.18	-0.10	-2.47 **	0.015
Regional dummies				
Middle East and North Africa	3.94	0.12	2.75 ***	0.007
Intercept	20.29		3.99 ***	0.000
Model statistics				
Number of observations	91			
R-squared, adjusted	0.884			
Standard error	3.623			
F statistic	86.800			

Source author's own calculation based on data described in Chapter 7.2

Notes 1 Diphtheria, Pertussis, Tetanus
2 refers to wars begun, ended or continued in 1983-92 or 1988-97 (matched with NI data)

Development Economics and Policy

Series edited by Franz Heidhues and Joachim von Braun

Band 1 Andrea Fadani: Agricultural Price Policy and Export and Food Production in Cameroon. A Farming Systems Analysis of Pricing Policies. The Case of Coffee-Based Farming Systems. 1999.

Band 2 Heike Michelsen: Auswirkungen der Währungsunion auf den Strukturanpassungsprozeß der Länder der afrikanischen Franc-Zone. 1995.

Band 3 Stephan Bea: Direktinvestitionen in Entwicklungsländern. Auswirkungen von Stabilisierungsmaßnahmen und Strukturreformen in Mexiko. 1995.

Band 4 Franz Heidhues / François Kamajou: Agricultural Policy Analysis – Proceedings of an International Seminar, held at the University of Dschang, Cameroon on May 26 and 27 1994, funded by the European Union under the Science and Technology Program (STD). 1996.

Band 5 Elke M. Förster: Protection or Liberalization? A Policy Analysis of the Korean Beef Sector. 1996.

Band 6 Gertrud Schrieder: The Role of Rural Finance for Food Security of the Poor in Cameroon. 1996.

Band 7 Nestor R. Ahoyo Adjovi: Economie des Systèmes de Production intégrant la Culture de Riz au Sud du Bénin: Potentialités, Contraintes et Perspectives. 1996.

Band 8 Jenny Müller: Income Distribution in the Agricultural Sector of Thailand. Empirical Analysis and Policy Options. 1996.

Band 9 Michael Brüntrup: Agricultural Price Policy and its Impact on Production, Income, Employment and the Adoption of Innovations. A Farming Systems Based Analysis of Cotton Policy in Northern Benin. 1997.

Band 10 Justin Bomda: Déterminants de l'Epargne et du Crédit, et leurs Implications pour le Développement du Système Financier Rural au Cameroun. 1998.

Band 11 John M. Msuya: Nutrition Improvement Projects in Tanzania: Implementation, Determinants of Performance, and Policy Implications. 1998.

Band 12 Andreas Neef: Auswirkungen von Bodenrechtswandel auf Ressourcennutzung und wirtschaftliches Verhalten von Kleinbauern in Niger und Benin. 1999.

Band 13 Susanna Wolf (ed.): The Future of EU-ACP Relations. 1999.

Band 14 Franz Heidhues / Gertrud Schrieder (eds.): Romania – Rural Finance in Transition Economies. 2000.

Band 15 Katinka Weinberger: Women's Participation. An Economic Analysis in Rural Chad and Pakistan. 2000.

Band 16 Christof Batzlen: Migration and Economic Development. Remittances and Investments in South Asia: A Case Study of Pakistan. 2000.

Band 17 Matin Qaim: Potential Impacts of Crop Biotechnology in Developing Countries. 2000.

Band 18 Jean Senahoun: Programmes d'ajustement structurel, sécurité alimentaire et durabilité agricole. Une approche d'analyse intégrée, appliquée au Bénin. 2001.

Band 19 Torsten Feldbrügge: Economics of Emergency Relief Management in Developing Countries. With Case Studies on Food Relief in Angola and Mozambique. 2001.

Band 20 Claudia Ringler: Optimal Allocation and Use of Water Resources in the Mekong River Basin: Multi-Country and Intersectoral Analyses. 2001.

Band 21 Arnim Kuhn: Handelskosten und regionale (Des-)Integration. Russlands Agrarmärkte in der Transformation. 2001.

Band 22 Ortrun Anne Gronski: Stock Markets and Economic Growth. Evidence from South Africa. 2001.

Band 23 Patrick Webb / Katinka Weinberger (eds.): Women Farmers. Enhancing Rights, Recognition and Productivity. 2001.

Band 24 Mingzhi Sheng: Lebensmittelkonsum und -konsumtrends in China. Eine empirische Analyse auf der Basis ökonometrischer Nachfragemodelle. 2002.

Band 25 Maria Iskandarani: Economics of Household Water Security in Jordan. 2002.

Band 26 Romeo Bertolini: Telecommunication Services in Sub-Saharan Africa. An Analysis of Access and Use in the Southern Volta Region in Ghana. 2002.

Band 27 Dietrich Müller-Falcke: Use and Impact of Information and Communication Technologies in Developing Countries' Small Businesses. Evidence from Indian Small Scale Industry. 2002.

Band 28 Wolfram Erhardt: Financial Markets for Small Enterprises in Urban and Rural Northern Thailand. Empirical Analysis on the Demand for and Supply of Financial Services, with Particular Emphasis on the Determinants of Credit Access and Borrower Transaction Costs. 2002.

Band 29 Wensheng Wang: The Impact of Information and Communication Technologies on Farm Households in China. 2002.

Band 30 Shyamal K. Chowdhury: Institutional and Welfare Aspects of the Provision and Use of Information and Communication Technologies in the Rural Areas of Bangladesh and Peru. 2002.

Band 31 Annette Luibrand: Transition in Vietnam. Impact of the Rural Reform Process on an Ethnic Minority. 2002.

Band 32 Felix Ankomah Asante: Economic Analysis of Decentralisation in Rural Ghana. 2003.

Band 33 Chodechai Suwanaporn: Determinants of Bank Lending in Thailand: An Empirical Examination for the Years 1992 to 1996. 2003.

Band 34 Abay Asfaw: Costs of Illness, Demand for Medical Care, and the Prospect of Community Health Insurance Schemes in the Rural Areas of Ethiopia. 2003.

Band 35 Gi-Soon Song: The Impact of Information and Communication Technologies (ICTs) on Rural Households. A Holistic Approach Applied to the Case of Lao People's Democratic Republic. 2003.

Band 36 Daniela Lohlein: An Economic Analysis of Public Good Provision in Rural Russia. The Case of Education and Health Care. 2003.

Band 37 Johannes Woelcke. Bio-Economics of Sustainable Land Management in Uganda. 2003.

Band 38 Susanne M. Ziemek: The Economics of Volunteer Labor Supply. An Application to Countries of a Different Development Level. 2003.

Band 39 Doris Wiesmann: An International Nutrition Index. Concept and Analyses of Food Insecurity and Undernutrition at Country Levels. 2004.

FSC
www.fsc.org
MIX
Papier | Fördert
gute Waldnutzung
FSC® C083411

Kolibri 360 GmbH
Ferdinand-Jühlke-Straße 7
99095 Erfurt, Deutschland
produktsicherheit@kolibri360.de

Druck:
CPI Druckdienstleistungen GmbH
im Auftrag der
Kolibri 360 GmbH
Ferdinand-Jühlke-Str. 7
99095 Erfurt